Educational Partnerships and the State

Additional praise for *Educational Partnerships and the State*:

"Franklin, Bloch, and Popkewitz's Educational Partnerships and the State could not be more timely. The volume significantly increases our knowledge of the meaning, development, and operation of educational partnerships. Franklin and his colleagues powerfully connect educational partnerships to questions of citizenship development and education for a democratic society. Franklin, Bloch, and Popkewitz, simply put, help us to better understand what educational partnerships are and why genuine and effective partnerships are so crucial for the future of democracy. I enjoyed the book and learned a great deal from it."
—Ira Harkavy, Associate Vice President and Director, Center for Community Partnerships, University of Pennsylvania

"This book provides rich insight into the nature of educational partnerships, which have emerged in recent years as part of a larger effort to reconcile fairness and markets by forging a Third Way between left and right. In the name of inclusion and equity, these partnerships seek to bring about a devolution of responsibility for education from the state to the civil society and the individual. However, the studies in this book show that, as government abandons the effort to promote equality and as the discourse of democracy becomes the discourse of accountability, individuals find themselves with more responsibility for educational outcomes and less control over educational processes. The authors suggest that, in the end, educational partnerships may leave us with an overburdened citizen and an irresponsible state."
—David F. Labaree, School of Education, Stanford University

Educational Partnerships and the State: The Paradoxes of Governing Schools, Children, and Families

Edited by
Barry M. Franklin, Marianne N. Bloch, and Thomas S. Popkewitz

EDUCATIONAL PARTNERSHIPS AND THE STATE
© Barry M. Franklin, Thomas S. Popkewitz, and Marrianne N. Bloch, 2003
Softcover reprint of the hardcover 1st edition 2003 978-1-4039-6128-0

All rights reserved. No part of this book may be used or reproduced in any manner whatsoever without written permission except in the case of brief quotations embodied in critical articles or reviews.

First published 2004 by
PALGRAVE MACMILLAN™
175 Fifth Avenue, New York, N.Y. 10010 and
Houndmills, Basingstoke, Hampshire, England RG21 6XS
Companies and representatives throughout the world

PALGRAVE MACMILLAN is the global academic imprint of the Palgrave Macmillan division of St. Martin's Press, LLC and of Palgrave Macmillan Ltd. Macmillan® is a registered trademark in the United States, United Kingdom and other countries. Palgrave is a registered trademark in the European Union and other countries.

ISBN 978-1-349-52646-8   ISBN 978-1-4039-8264-3 (eBook)
DOI 10.1057/9781403982643

Library of Congress Cataloging-in-Publication Data
    Educational partnerships and the state : the paradoxes of governing schools, children, and families / edited by Barry M. Franklin, Marrianne N. Bloch, Thomas S. Popkewitz.
        p. cm.
    Includes bibliographical references and index.

    1. Community and school—United States. 2. Industry and education—United States. 3. Education and the state—United States. I. Franklin, Barry M. II. Bloch, Marrianne N. III. Popkewitz, Thomas S.

LC215.E27 2003
371.19′5′0973—dc21   2003051932

A catalogue record for this book is available from the British Library.

Design by Newgen Imaging Systems (P) Ltd., Chennai, India.

First edition: December 2003
10 9 8 7 6 5 4 3 2 1

Transferred to Digital Printing 2009

Contents

Acknowledgments — xi
About the Contributors — xii

Introduction: Educational Partnerships: An Introductory Framework — 1
Barry M. Franklin, Marianne N. Bloch, and Thomas S. Popkewitz

Part I Partnerships and Changing Patterns of Governing

Chapter 1 Partnerships, the Social Pact and Changing Systems of Reason in a Comparative Perspective — 27
Thomas S. Popkewitz

Chapter 2 Partnership as a Floating and Empty Signifier within Educational Policies: The Mexican Case — 55
R. Buenfil Burgos

Part II Between the State, Civil Society, and Education

Chapter 3 Partnerships in a "Cold Climate": The Case of Britain — 83
Barry M. Franklin and Gary McCulloch

Chapter 4 Education Action Zones: Model Partnerships? — 109
Marny Dickson, Sharon Gewirtz, David Halpin, Sally Power, and Geoff Whitty

Chapter 5 Partnerships: The Community Context in Miami — 137
M. Yvette Baber and Kathryn M. Borman with Jennifer Avery and Edgar Amador

Part III School Reform and Public–Private Partnerships

Chapter 6 The Public–Private Nexus in Education 171
Henry M. Levin

Chapter 7 Governance and Accountability in the Michigan Partnership for New Education: Reconstructing Democratic Participation 187
Lynn Fendler

Part IV Governing as a Problem of Inclusion and Exclusion

Chapter 8 Partnerships and Parents: Issues of Sex and Gender in Policy and Practice 213
Miriam E. David

Chapter 9 Partnering to Serve and Save the Child with *Potential*: Reexamining Salvation Narratives within One University–School–Community "Service Learning" Project 237
Marianne N. Bloch, I-Fang Lee, and Ruth L. Peach

Index 267

Acknowledgments

We would like to thank a number of individuals who were instrumental in the preparation of this volume. Peggie Clelland, a doctoral student in curriculum and instruction in the Department of Secondary Education at Utah State University, played a major role in proofreading the entire manuscript. Chris Kruger, a secretary in the Department of Curriculum and Instruction at the University of Wisconsin-Madison, assumed responsibility for the typing and formatting of several chapters. The entire volume received a critical reading from the participants of the Wednesday and Thursday graduate student seminars in the Department of Curriculum and Instruction at Wisconsin. Their comments were especially helpful in offering suggestions for revisions to our contributors. Our editor at Palgrave Macmillan, Amanda Johnson, championed this volume from the beginning and offered support and encouragement all along the way. Finally, we would like to thank the MIT Press for permission to reproduce Henry Levin's chapter, which originally appeared in Pauline Rosenau (Ed.), *Public-Private Policy Partnerships* (Cambridge: MIT Press, 2000), pp. 129–142.

About the Contributors

Edgar Amador is a graduate student in applied anthropology at the University of South Florida.

Jennifer Avery is a graduate student in applied anthropology at the University of South Florida.

M. Yvett Baber is an independent researcher in applied anthropology.

Marianne N. Bloch is a Professor of Curriculum and Instruction at the University of Wisconsin-Madison.

Kathryn M. Borman is a Professor of Anthropology and Assistant Director of the David C. Anchin Center at the University of South Florida.

R. Buenfil Burgos is a Professor of Education at Cinvastev (Mexico).

Miram E. David is a Professor of Policy Studies in the Department of Education and Director of the Graduate School of Social Sciences at Keele University.

Marny Dickson is a Research Officer in the Education Policy Research Unit at the Institute of Education, University of London.

Lynn Fendler is an Assistant Professor of Education in the Department of Teacher Education at Michigan State University.

Barry M. Franklin is a Professor and Head of the Department Secondary Education at Utah State University.

Sharon Gewirtz is a Professor of Education at King's College, University of London.

David Halpin is a Professor of Education and Head of the School of Curriculum, Pedagogy and Assessment at the Institute of Education, University of London.

I-Fang Lee is a doctoral student in curriculum and instruction at the University of Wisconsin-Madison.

Henry M. Levin is the William H. Kilpatrick Professor of Economics and Education at Teachers College, Columbia University.

Gary McCulloch is the Brian Simon Professor of the History of Education at the Institute of Education, University of London.

Ruth L. Peach is a doctoral student in curriculum and instruction at the University of Wisconsin-Madison.

Thomas S. Popkewitz is a Professor of Curriculum and Instruction at the University of Wisconsin-Madison.

Sally Power is a Professor of Education and Head of the School of Educational Foundations and Policy Studies at the Institute of Education, University of London.

Geoff Whitty is the Director of the Institute of Education, University of London.

Introduction
Educational Partnerships: An Introductory Framework

Barry M. Franklin, Marianne N. Bloch, and Thomas S. Popkewitz

In his address to Britain's New Labour Party Conference in October 2002, former U.S. president, Bill Clinton, offered his recipe for a more secure, more prosperous, and ultimately a "more integrated global community." Such a world, he reminded his audience, required the extension of human rights to all peoples, support for international agencies that promote peace and security, and efforts to combat the spread of nuclear, chemical, and biological weapons. Most importantly, he went on to say, is that such a future required that we "build a world with more partners and fewer terrorists" (Clinton, 2002, p. 6). A day earlier, Prime Minister Tony Blair in his address to the Conference had sounded something of a similar theme. "Partnership," he noted "is the antidote to unilateralism." "Partnership is statesmanship for the 21st century." And "partnership is also citizenship for the 21st century." Partnerships, as Blair sees them, are the vehicles for the delivery of the array of services and provisions that contemporary society typically offers its citizens (Blair, 2002, pp. 3, 7, 17).

Clinton and Blair are not alone in invoking the idea of partnerships as a political discourse about the revitalization of democracy. Conservative politicians have been equally enamored of the notion. Most recently, George W. Bush has framed his proposal for faith-based initiatives using the discourse of partnerships ("President Bush's Faith-Based & Community Initiatives," n.d.). In their essay (chapter 3) in this volume, Franklin and McCulloch note that when the Conservatives came to power under Margaret Thatcher in Great Britain in 1979, they looked to the United States for possible ideas for educational reform. Attracted by the school-business compacts that they saw when several government ministers visited America, they established a number of partnership initiatives, including the City Technology College program. And a number of left-of-center leaders in Europe and the rest of the world, often under the guidance of international agencies' requirements to reduce public spending, have also promoted partnerships as a reform strategy.

Many of these political supporters of partnerships are associated with a current reform movement known as the Third Way. They advocate public–private collaborations as a reconstructed view of the state that does not fall nicely into either of the two paradigms that have dominated democratic political thinking since the end of World War II. That is, it is a viewpoint that does not fit the kind of social democratic thinking that prevails among those who lean leftward politically. And similarly, it is, at the same time, an understanding of the state that challenges much of the market-oriented programs and policies associated with neoliberalism (Blair, 1998; Coates, 2000; Giddens, 1998; Latham, 2001; Muravchik, 2002). However, despite the flourish that a label like the Third Way gives to this movement, it is actually not all that new or unique. What it represents in fact is the most recent version of a long-standing effort of liberal governments to recalibrate the relationship of governing through the overlapping practices that link civil society and the state. The changes occurring are not merely a recurrent historical pattern but involve particular sets of relations that require systematic examination.

The crossing of partnerships between the ideological agendas of neoliberalism and the Third Way enables us to think about the changes in the contemporary landscape of governing where ideological concepts no longer provide explanatory power. The policies of partnerships are both a reform being implemented and expressive of a more general comparative concern with the changing patterns of governing that are embodied in such reforms. The reforms that occur under the rubric of partnerships are ways to engage in a systematic inquiry into the changing governing patterns that link the state and civil society. Here, the notion of a partnership provides a way of understanding the regulative mechanism of society in which the state as entity is only one of a number of sectors that include business, religion, voluntary agencies, families, and individuals themselves in the governing of society and its citizenry (see, e.g., Rose, 2000; Wagner, 1994). At the same time, it offers a means of exploring political rationalities of governing in the ordering of the inner motivation and responsibilities of the child, family, and community (Bloch et al., in press).

The volume explores partnerships in educational reform as not merely organizational processes to bring people together but as practices in reconstructing the contemporary state and social/cultural relationships through which the qualities of the citizen are produced. Taken together, the chapters comprising this volume suggest that such reforms bring with them an array of paradoxes in the regulation of schools, children, and families—central sites of controlling and constructing the characteristics and capabilities of the individual who acts and participates in civil society. Among these are efforts for social inclusion that actually work to create systems of exclusion, attempts to expand civil society that ultimately serve to enhance the regulative power of the state, the practices of individual empowerment that in fact diminish such popular control, and efforts for local involvement and participation that emerge as elements of global discourses of state and interstate policy. It is for these reasons that we entitled this collection of essays as *Educational Partnerships and the State: Paradoxes of Governing Schools, Children, and the Family.*

Meanings of Partnership as it Enters into Today's Educational Reform Discourses

Partnerships describe the efforts of groups and individuals including government, business, the church, the voluntary sector, and parents, to work collaboratively to solve problems, in this case educational problems (Blair, 1997; Commission on Public Private Partnerships, 2001; Maurrasse, 2001). Such arrangements are networks that establish patterns of association and interaction that link the state to civil society with the intent of forming interconnections, introducing flexibility, and structuring individuality and citizenship. Used in this way, partnerships appear to attenuate a host of once important distinctions between public and private, left and right, government and industry, state and market. With respect to schooling, partnerships represent one aspect of new patterns of governing, and of potentially organizing the participation and experience of those often left out of policy discussions or decision-making, theoretically bringing new "voices" to the partnership table. The new voices range from families or communities left out of past reforms, to churches or other religious organizations, to nonprofit and for-profit corporations. Levin's chapter (chapter 6) in this volume, for example, illustrates a number of forms that these educational partnerships have taken.

Partnerships can be thought of as an example of what Antonio Nóvoa (2002) calls "planetspeak," his term for an almost magical concept that seems to offer the solutions to all problems while at the same time rooting out all evil. Planetspeak, according to Nóvoa, brings with it a new expert who creates and circulates international discourses that seem to exist without structural roots or social locations. He views such discourses as a "worldwide bible" whose vocabulary serve as banalities universally accepted as truths that have no known origin and do not need to be questioned. Invoking the notion of a partnership seems, then, to suggest that there exists a consensus among the various parties charged with the solution of any problem. Buenfil's essay (chapter 2) on partnerships in twentieth-century Mexican reform discourse in this volume is illustrative. As she sees it, partnerships are what she calls "floating and empty signifiers" whose very ambiguity allows for an array of responses to the dilemmas facing Mexican state schooling, including but not limited to political action, hegemonic practices, dissention, and negotiations. It is, she goes on to say, the very emptiness of the concept that gives it the universality that accounts for its wide usage in educational and other reform discourse.

What seems in modern parlance to distinguish a partnership from other cooperative or collaborative arrangements is the extent to which its members appear to share responsibilities and risks associated with their joint venture (Linder and Rosenau, 2000). Partnerships are to have joint goals/risks/responsibilities in forming visions and strategies in social projects. The goal of these partnerships often appears in the formation of networks and alliances within civil society—seemingly outside the control of the market or the state.

Partnerships: A Reexamination of the Newness of Reform Discourse

Clearly the idea of partnerships as reform enterprises is not new in the educational arena. It is embodied in modern political theory that establishes a relation between the state and civil society from Locke to recent American pluralistic theories of the middle of the twentieth century. If we look at the early twentieth-century efforts to expand U.S. schooling to include kindergartens, vocational schools, schools for exceptional children, and community schools, they were often promoted through cooperative arrangements between the state and an array of charitable and voluntary organizations (Franklin, 1994; Kliebard, 1999; Reese, 2002). Similarly, formal linkages between parents and schools began to appear in the last quarter of the nineteenth century (Cutler, 2000). These ventures can be thought of as the earliest examples of what we today label public–private and school–parent partnerships.

During the 1960s partnerships began to assume increasing importance in the reform of education as part of U.S Federal and state initiatives to reduce poverty and improve the academic performance of racial minorities and the urban poor. Four of the most prominent of these programs were the Job Corps, Project Head Start, Teacher Corps, and performance contracting. The first involved a number of the nation's largest corporations, including IBM, General Electric, and Westinghouse, in the operation of residential centers offering basic education and vocational training to unemployed and undereducated youth. Project Head Start involved a partnership between local communities, parents, social service agencies, and preschools to provide young disadvantaged children with early education and other social services while at the same time offering parent education to their caregivers. Teacher Corps represented a collaborative between universities, the Federal government, and local communities to provide more teachers for disadvantaged communities while at the same time enhancing the quality of their training.

Finally, performance contracting represented an agreement to enable a private firm to manage or provide services and materials to one or more schools within a district in return for guaranteeing improvements in managerial effectiveness and student achievement. These were schemes in which the fees that these firms were able to charge school systems as well as the continuation of their contracts hinged on their ability to achieve agreed-upon performance goals (Campbell and Lorion, 1972; Levitan, 1969; Mecklenburger, 1972; Silver and Silver, 1991). One of the key benefits of these partnership ventures, which were continued during the Nixon, Carter, and Reagan administrations, was their ability to secure private sector financial support for an array of public services (Berger, 1985).

These partnerships captured a political agenda to bring the poor into the decision-making apparatus of social and state institutions. In practice, however, such programs emphasized administrative procedures to gain greater parent and community support of institutional goals (Popkewitz, 1976). The decision-making processes of the partnerships emphasized particular norms of behavior and action of the existing cultures of the curriculum and education at school as well as in homes and communities. There was also a psychological focus in the programs that sought to increase the political efficacy and self-esteem of those affected by poverty through involvement in decisions that affected their lives, and through on-the-job training and parent education.

The history of partnerships is not an evolutionary story of the broadening participation in the processes of the schools and society. The linkages between parents, communities, and schools in the nineteenth century involved particular historical trajectories related to governing the school, children, and parents. Rose (1999) argues that the family, community, and the school in Western Europe and the United States were places where the notions of freedom and liberty were problematized and reordered through political theories and pedagogical programs. The local and communal involvements were strategies that functioned to reconstitute and shape the norms of public duty inscribed in the family while not destroying the private authority of the family.

Wagner also suggests that notions such as partnerships require studying the relations of state administration and the freedom of the individual. He argues, "if modern institutions do not merely enhance liberty but offer a specific relation of enablement and constraint, the substance and the distribution of enablements and constraints become important. The history of modernity cannot simply be written in terms of increasing autonomy and democracy, but rather in terms of changing notions of the substantive foundations of a self-realization and of shifting emphases between individualized enablements and public/collective capabilities" (Wagner, 1994, p. xiv). Today's partnership reforms involve this double relation of linking the governing patterns of the state with civil society *and* the principles of individual action. This relation is found throughout the chapters of the book. But there is a different amalgamation of cultural and social practices embodied in the partnerships that this volume also seeks to illuminate (see Popkewitz, 2003).

Globalization and Localization

The renewed interest in partnerships that now appears is related in public discourse to an array of social and economic changes collectively labeled as globalization. Most important of these changes has been the emergence of worldwide communication and transportation systems. Other changes, all of which require global communication and transportation networks, include the expansion of trade and investment across national borders, the growth of international financial markets, the immediate impact of distant events, the global effect of local economic and political decisions, and the redistribution of power among nation-states, local entities within nations, and within regions that span national boundaries. Taken together these changes are viewed in policy as producing a world that is less stable, less secure, and more subject to immediate change (Giddens, 1990, 1994; Gilpin, 2000; Mittelman, 2000; Reich, 1991; Rose, 1999). Partnerships, for their proponents, represent a policy initiative for coping with these new conditions. Seen as an antidote for historic battles between left and right and public and private, partnerships are to bring government, business, the voluntary sector, and citizens together around issues of public policy. It is in other words—in appearance—a mechanism for cooperation and consensus.

At the same time, partnerships are alternatives to the regulative role neoliberals bestow on the market without falling back on the statist policies of social democracy. Partnerships in fact appear to offer a way for those on the left to endow the state with a positive role to play in facilitating fairness, equality, and social inclusion at a time when the state itself often lacks the financial resources and the political will to attain these goals (Blair, 1997; Hatcher and Leblond, 2001), and through the guise of decentralization, often reduces rather than increases its funding for educational and other social programs.

The claims of globalization involve a complex set of historical relations between the global and the local that needs scrutiny. First, globalization is not a new phenomenon as it was found, for example, in the early Catholic Church, the spread of Sanskrit, and later with the worldwide spread of the nation-state and the school (see Meyer, Boli and Ramirez, 1997). Our inquiries in this book, then, view globalization as a concept related to distinct historical patterns that simultaneously involve the national, regional, local, and global.

Second, while we view partnerships as part of a globalization of economic, political, and cultural knowledge and relationships, the essays are not concerned with measuring partnerships as an absolute evaluation of a program's goodness or badness. Rather, ours is an empirical and comparative inquiry into how the practices of educational reforms known as "partnerships" are constituted in an array of social changes.

Third, we view partnerships as historical responses to and embodying particular social dislocations, anxieties, and tensions produced through such factors as immigration. There has been an enormous movement among populations, with migrations across different countries and regions, that has been an important part of the processes of globalization. These demographic changes are represented, for example, in the emergence of multicultural programs in the United States and the European Union. In chapter 5, Baber and Borman look at two senior high schools in the greater Miami area to explore the interplay between issues of immigration and the use of partnerships to enhance student achievement and opportunity.

But the migratory patterns are not only across national borders. There also have been dramatic movements from rural to urban locations in the United States and Europe since the second half of the twentieth century. These demographic changes have produced new expressions in national policies about moral disintegration, instability, and lack of a consensus as previously thought of homogeneous nations become fearful of their future.[1] While the ideas of the homogeneous nation may have never been accurate, the policies of partnerships serve a double function. They appear as inclusive strategies for those who have not had access to public institutions. And the policies of representation and democratization in partnerships provide a setting for creating patterns of stability and moral cohesion in the contexts of ongoing cultural and social uncertainty.

Partnerships are, however, not just policy initiatives in the realm of governing. In this volume they are considered as a field of cultural practices that is embedded with specific constructions of democracy and citizenship. Partnerships in other words presume a certain type of state, one that promotes the devolution of authority and the decentralization of decision-making. This is accomplished through initiatives that increase the scope and involvement of civil society through the engagement of nongovernmental groups and agencies. This increased involvement and its tensions and conflicts are explored in chapter 4 by Dickson, Gewirtz, Halpin, Power, and Whitty in this volume. A complex relation of state mandates to promote new networks in local decisions-making are explored as practices that do not always meet up to their expectations and involvement.

Chapter 1 by Popkewitz, in contrast, explores how partnerships are embedded in a field of cultural practices that reconstitute the governing processes of schools. They represent one set of practices in an array of changes in the organization of schools as well as in curriculum and pedagogy that link the collective rationalities of the acting subject with the construction of principles that govern individuality.

This does not mean, however, that governing does not occur through a relation of the center and the local. If we take neoliberalism, for example, it involves policies used in different places with different political and cultural agendas that seem at first glance as strange bedfellows. Russian strategies of marketization are embodied in a field of social and political practices that include a skillful, once Communist apparatus that moves within the shadow of strong state governing that has little concern for a democratic society agenda. Argentinean reforms of neoliberalism involve current practices of restructuring that embody a double technology that both empowers and disables—includes and excludes (Dussel, in press). To understand the double technology is to investigate in a combinatory repertoire that relates to new patterns of governance based on personal responsibility and self-government.

In effect, the intimate participation of the state in the lives of its citizens appears to be replaced by a more distant form of governing. However, partnerships and networks also embody a notion of the state as the conditions through which the political rationalities of governing are brought into a practical relationship with the cultural practices through which the citizen is produced. This notion of the state draws on Foucault's idea of governmentality in which the micro practices of governing society are linked with the patterns of decision-making and rules of reasoning by generating the principles of individual action and participation. This view of governing directs attention to how current reforms to decentralize and to provide greater community and individual involvement that flow so easily as "planetspeak" in contemporary reforms are not in opposition to the problem of regulating but are very much part of the governing patterns.

Citizenship under this brand of liberal democracy from the nineteenth century onward sees the relationship of the individual to the state as a partnership built on mutuality and reciprocity. It is a relationship predicated on the notion of "no rights without responsibilities" (Giddens, 1998, p. 65). In other words, social provision, particularly unemployment benefits, welfare payments, and health services that are part of a social safety net but also education and training, are no longer seen as simple entitlements, as they have been viewed by some social democratic regimes. Rather, increasingly, in welfare reforms throughout the world, social supports entail personal responsibility and self-sufficiency on the part of individuals to actively seek and accept paid work and to commit themselves to a process of lifelong learning that will maintain and strengthen their economic viability and, supposed, autonomy from the state. The individual is to take added responsibility for governing with the rules and boundaries inscribed by the narratives of responsibility (Foucault, 1991; Rose, 2000).

Fourth, the society emerging from this new mode of governing is fashioned in a way that appears to be more equal and more socially inclusive. In part, this involves the cultivation of a civic republicanism in which individuals are active participants in politics and public affairs. It also, however, involves an acceptance of the obligation of citizens to try to maintain their economic viability through employment and training, a particularly difficult task for many especially during a period of high unemployment. This new societal organization, locally, and globally, therefore, involves new patterns of exclusion that are masked in the language of inclusion—of needing to be responsible or autonomous (Levitas, 1998).

Social Inclusion/Exclusion as a Problematic of Government and Governing

This pattern of inclusion and exclusion mediated through partnerships is best illustrated in the way in which many New Labourites in Britain talk about the problem of equality. Their view of equality stands in contrast to existing social democratic views that equate the concept of equality with that of equalizing the condition of life among the citizenry through such policies as increased public ownership, a redistributive taxation scheme, an extensive system of public welfare, and a commitment to full employment (Heffernan, 2000; Howell, 1980; Laybourn, 2000; Levitas, 1998).

Equality means something different, namely equality of opportunity. They explicitly reject redistributive policies that seek to reduce the economic differences that separate the rich from the poor in favor of efforts that they claim will enhance everyone's chances to succeed, including emphases on personal or individual responsibility and motivation, incentives, entrepreneurialism, and enhanced human capital development through education (Giddens, 1998). Invoking notions of community, they envision a society in which all its citizens have a share in the fate of their nation and are bound together in a relationship of what Tony Blair has referred to as "social unity, common purpose, fairness, and mutual responsibility" (Blair, 1997, p. 299). It is an approach to governing that its promoters see as avoiding what is seen as both the inflexibility and economic inefficiencies of social democratic statism without falling prey to the vicissitudes that they believe have accompanied the faith that neoliberals have in free-market philosophies and strategies (Blair, 1998).

The route to equality is through social inclusion, which in their lexicon refers to access to education and work. However, this is not considered so much as a right as it is an obligation. In return for their ability to have access to schooling, training, employment, and to participate in democratic society, individuals have the responsibility to look for and obtain paid employment, and to acquire required skills, knowledge, and conduct that they appear to need to be prepared for work and to participate in society (see Fraser, 1997; Schramm, 2000 for counterarguments). It is not, however, only the earnings that make work inclusive, according to this reasoning. Rather it is also the sense of self-esteem and self-worth that being employed holds out for individuals (Levitas, 1998).

The importance in this discourse shift becomes clear when we look at how it plays itself out in policy, specifically educational policy. Take the central educational problem of most contemporary Western democracies, educational underperformance. Third Way thinkers pay scant attention to the structural problems that social democrats typically see as the culprits, namely economic inequality, class or gender bias, or racism. Like many neoliberals they see it as a problem of deficient families, especially parents whose own history of unemployment, poverty, and lack of education has left them supposedly without the skills, self-esteem, and ambition to help their children succeed in school. This is the same kind of reasoning that labels single, working mothers as deficient and irresponsible because the demands of full-time employment preclude their ability to become involved in their children's schooling. It is also a viewpoint of poor and minority parents that is the impetus for the creation of parent education as part of the Head Start Program.

Like many neoliberals, Third Way proponents favor policies such as home-school contracts, parent education, and mandatory curfews that virtually compel parents to assume responsibility for their children's social behavior and academic success (Arthur, 2000). Miriam David's chapter in this volume (chapter 8), for example, discusses these policies in relation to gender and generation in the contexts of Great Britain and the United States, particularly as neoliberal policies relate to the moral panic that occurs in global, familial, and social transformations and the revisions to notions of sex and sexual identities. Where they differ from neoliberals is in their advocacy of policies that offer parents greater access to education and training, which they need to obtain employment, and which, in turn, they claim, will provide them with the desire and sense of self-worth to engage themselves in their children's lives and schooling.

For proponents of the Third Way, the strategy of choice in education was the establishment of partnerships within civil society. Parents are to enter into partnerships with the state to provide them with the knowledge, dispositions, and skills that will make them employable *and* fabricate them as good parents for their children and for the democratic, increasingly cosmopolitan, and globalized nation and world. They are also to enter into partnerships with the schools to "involve" themselves in multiple ways and work toward their children's academic success (Bloch and Popkewitz, 2000; Bloch, Holmlund, Moqvist and Popkewitz, 2003; Gamarnikow and Green, 1999; Gewirtz, 1999;

Gillborn, 1998; Power and Gewirtz, 2001; Power and Whitty, 1999). For Baber and Borman (chapter 5), the role that parents play in partnership initiatives is linked to class. They argue in their study that middle-class parents were much better positioned than lower-class parents to help their children take advantage of what community partnerships had to offer in promoting opportunity. Among the working class, the family rather than the community was the more important vehicle for support for children. This, then, affects the way that, according to David in chapter 8, gender relations influence policies to provide boundaries in which the family makes free choices and has opportunities.

The term equity is used in neoliberal and Third Way policies to express a concern with social inclusion and exclusion. Equity is linked with policies of partnership as a means of finding solutions to identifiable problems. But as with the term partnership, equity is not merely a word that expresses normative commitments. It is what Ian Hacking calls an *elevator word*, a word that appears as a "fact," "truth," and "reality" but that is circularly defined and free floating so as to seem uncontroversial and thus in need of no definition (Hacking, 1999, pp. 22–23).

In this book, we place such elevator words as equity within a series of different national and educational practices that constitute partnerships. Our strategy of inquiry is to consider different *problematics* that order and define partnerships. By problematic, we mean the relation of methods, concepts, and theories of social affairs in constructing ways of thinking and ordering action and of conceiving of results or effects.

The equity problematic. One such problematic that we discussed earlier is the one concerned with *equity*. To be schematic, a concern with equity expresses a particular philosophical and social set of ideas about inclusion that emerges with what can be thought of as the social question of the nineteenth century. This involves a belief that rational action and a collective authority can produce the expectations and entitlements of individuals who act as agents of their own interests (see, e.g., Giddens, 1990; Wagner, 1994). Problems of policy and research about partnerships are to eliminate exclusions by identifying the best paths for social mobility, access, and social and cultural representation of groups. The empowerment of marginalized groups to participate in local school councils would be one example of the equity problematic. Providing children and families greater choice to attend private schools, through a public–private funding partnership, thereby improving educational opportunity and access to presumably higher quality schools, would be another.

Nineteenth-century social thinkers faced a double problem when they confronted this problematic. The concern of social policy was not only how to humanize and bring progress to society in the face of the destabilizing changes of industrialization and urbanization. There were also salvation themes about being able to rationalize and give order and progress to society (Comte, 1827/1975). The focus of policy and social science was on rational action and a collective authority that enabled individuals to act as agents of their own interests in the imagined interests of the nation. Today's social question involves fears of social–cultural disintegration and a moral disorganization that notions of globalization and concerns about nation and cultural identity articulate. The equity problematic in today's policy and research embodies the assumptions of the governance posed by the questions of equity, social inclusion/exclusion, but with different sets of nuances and institutional structures for seeking of greater social mobility, access, and the social representation of groups.

Partnership as a salvation theme is explored in chapter 7 by Bloch, Lee, and Peach as well as in chapter 1 by Popkewitz. Salvation narratives of educational reforms speak of the state's collective obligation to promote community involvement that empowers previously marginalized groups. Traveling through the different ideological scenarios is the professional teacher who revives and reinvents democracy by working more directly with parents and communities through, for example, partnerships.

Knowledge/power relations as systems of reason about social inclusion/exclusion. Alongside but different from the equity problematic in this book is a problematic that focuses on the knowledge or systems of reason that link issues of equity with partnerships. Knowledge is viewed culturally and historically to consider how objects of reflections and principles for action are ordered (the two problematics are discussed in Popkewitz and Lindblad, 2000). Whereas the equity problematic treats the knowledge of schooling as the representing of social interests, the knowledge problematic focuses on the rules and standards of systems of reason.

It is at this point, we can think about partnerships as practices of inclusion and exclusion in ways that are somewhat different from that of the equity problematic. The equity problematic locates the politics of governance as identifying processes that include or exclude certain groups from participation. The knowledge problematic, on the other hand, focuses on the principles of classification that make possible the range of practices and possibilities considered and who is signified as the reasonable person in these processes of change.

The ordering procedures to define reason and the reasonable person not only provide a way to consider how knowledge functions to qualify and disqualify individuals for participation. Butler (1993) argues that uncovering of opposition, of hierarchies and orders of subordination is to expose the artifices and exclusions inherent in the categories of nature, gender, class, and citizen. This turn to knowledge as a social practice is, as Butler argues, to question the constitution of the subject of modernity as a particular invention of Western philosophy. Butler continues that when the subject is taken uncritically as the locus of struggle for knowledge about enfranchisement and democracy, scholarship draws from the very models that have oppressed through the regulation and production of subjects. Such a strategy is both a consolidation and concealment of those power relations.

The distinctions between the two problematics enable a consideration of different analytics of inclusion and exclusion in this volume. The equity problematic makes an analytic distinction between inclusion and exclusion. Policy and research investigate how policies and practices are reducing or eliminating exclusion. The theoretical possibility of this approach is that the correct policies will enable, eventually, a totally inclusive society. The knowledge problematic focuses on inclusion and exclusion as one concept (inclusion/exclusion). This second side enables us to consider partnerships in which issues of exclusion are continually placed against the background of something simultaneously included (Goodwin, 1996). Levin's treatment of educational vouchers in his chapter offers a good example of how these two problematics come into play in considering the role of one kind of partnership enterprise as a vehicle of reform.

In this volume, the contributors illustrate both perspectives through case studies of different educational partnerships. In their exploration of Education Action Zones in this volume, Dickson, Gewirtz, Halpin, Power, and Whitty (chapter 4) explore how a reform that ostensibly promotes inclusion actually serves to reproduce the very inequities that it was designed to combat. Fendler's essay (chapter 7) illustrates these two perspectives by focusing on the reasoning of "communities" as these are part of the professional development school reform discourse in Michigan. The concern is with how the objects of the world, self, and community are known through historical processes. In examining the reforms related to partnership, her research, as in the research of other contributors, is interested in the classifications and differentiations that form the objects of the partnerships. The significance of the rules and standards of reason of partnerships is that they discipline, normalize, and make possible certain ways of thinking and acting. This notion of inclusion/exclusion is treated in Bloch, Lee, and Peach's essay (chapter 9) on student volunteerism and service learning. The reasoning embedded in the university–school partnership they describe constructs the students served in these schools as having potential, which obscures another pattern of reasoning in the program, namely that without university intervention these students are constructed as at-risk for failure. They argue that partnerships open up a "third space," between the state and civil society in the governing of individual. It is in that space that the production of principles are created that qualify and disqualify individuals for of action and participation.

Partnerships, then, are approached in this volume from at least these two different problematics related to social inclusion/exclusion. Engaging the two problematics in a conversation with each other is important for an adequate understanding of contemporary politics as well as for being reflexive about the politics of knowledge in science. Not to recognize the complex relations of governance and inclusion/exclusion in the projects of partnership is to lose sight of the politics of intellectual and social projects. Not to recognize the different ways in which theories frame our modes of analyzing reforms, their problematics, and results/effects, and possibilities would reduce our ways of imagining possibilities in the end.

As salvation narratives, partnerships embody a distinct view of progress as reinstalling the religious ideas of redemption in the secularization produced by the new political regimes of democracy and the economic forms of globalization. Instead of the Christian view of salvation as occurring in the afterlife, human fulfillment in Third Way thinking occurs through modern welfare institutions and philosophies that tie individual betterment to social progress. The scholarship on world systems, for example, ties the development of the school to salvation narratives that link the nation with individual actors as an agency of change that appear to fabricate a better self, family, and nation. Therefore, one can think of the current notion of partnership as not only reestablishing relations between the regulatory functions of the state with civil society, but also as offering new salvation themes of individual and national redemption for a globalized world (Popkewitz, 1998; Popkewitz and Bloch, 2001).

At the heart of this volume, then, is a critical dialogue that makes educational partnerships into a site to investigate broader changes in the governing patterns that organize schooling and the notions of citizenship that such governing embodies. At one level, we ask about partnerships as strategies to address educational problems including persistent low achievement, declining educational standards, the gap between schooling and employment, and the apparent inability of schools to promote civic engagement. At a different level, the contributors to this volume ask about the different social and cultural practices embodied in partnerships.

Governance as explored by the contributors to this volume has a dual quality. One involves viewing partnerships as social and administrative practices that promote or limit integration or access of social groups and individuals. This notion of governance is typically related to the formal administrative practices of the state. *Governing* is to set in place the proper rules and procedures to organize institutional practices that increase the inclusion of different populations. But as we argued in the earlier discussion of the equity and knowledge problematics, the governance of partnerships can also be viewed through examining the knowledge or systems of reason that orders the objects of the world and self to be known and acted on. Governance, in this second sense, is the practical rationalities of daily life, know-how, expertise and means of calculation that structures the field of possible actions and participation. In light of these concerns with partnerships and the governance of education, the essays address a number of important questions.

First, how do the mechanisms of partnerships relate to questions of governing and who governs? Are partnerships mechanisms for bringing an array of parties concerned about a particular issue to the same table? Whether we are talking about issues of trade, immigration, international relations, or schooling in today's world, it is typically unclear exactly what kind of policies are required to address the issue at hand and who should participate in shaping those policies. Partnerships are often promoted on the grounds that they serve, so to speak, to cover all the possible bases by bringing together all the players and all the possible solutions. Uncertainty is, ironically, the hallmark of today's pressing educational dilemmas. We are not sure what kind of policies are required to resolve the problem of inequities and exclusions, nor are we clear about who the key players should be in formulating those policies. Do partnerships allow us to enlist the participation of an array of individuals and groups in dealing with the multiple causes and varied problems associated with low school achievement, or do they exclude in a variety of ways?

Second, how successful are partnerships in reconciling the contradictory forces that impinge on contemporary life? Partnerships as strategies find support from both neoliberal policies that see markets as the key instrument for the allocation of social provisions and benefits. At the same time partnerships are promoted by Third Way principles of equality and fairness. In this environment, can partnerships serve to harmonize markets with fairness? Can they bring together a number of competing individuals and groups so that those concerned with equity can play a part in the workings of the market? In what ways do we reason about the ethics involved with different cultural and political conceptions of "fairness"?

Third, what is the nature and character of civic engagement and democracy promoted through partnerships in different state traditions? One of the central concerns of today's policymakers is that ordinary citizens are increasingly excluded from the important decisions of political life. Partnerships are often touted as devices that, at least theoretically, make possible the coming together in one forum the most and least powerful members of society. The conception of bringing the experiences and voices of those who have previously been excluded is an important aspect of the democratization and enhancement of citizenship that provide a rationale for home-school collaboratives, enhanced parent involvement, local community councils, site-based management, and even some faith-based initiatives. In other words, how do educational partnerships function in relation to different stakeholders in the system and in relation to unequal power relations? Whose "voices,"—students, teachers, communities, or parents—are being targeted to be heard as "involved" as partners?

Fourth, what are the territories or space created for addressing societal problems produced through the dynamics of partnerships? One of the most important features of partnerships, according to many proponents, is that they represent a new venue for problem-solving where a commitment to a sense of community governs. They are said, then, to offer an antidote for the divisiveness and self-interest that dominates contemporary politics and policymaking. To what extent, however, are partnerships guided by a commitment to the common good and a goal of common purpose? How is this defined, when the very notion of democracy, and "common purpose" defies the multiplicity of possibilities that are available, and/or have been hidden from view through the reasoning of what we assume as natural, science, normal, or good must be questioned, and opened to new possibilities.

Fifth, what are the patterns of governing produced and embodied in the practices of partnerships? How do different sets of ideas, institutions, stories, and theories of the child/parent/teacher circulate in the spaces of partnerships? What governing principles circulate in the relationships and networks? What are the relationships of civil society and the state inscribed through the practices of partnerships? And how do partnerships help us understand changing patterns of power?

And sixth, how can the notion of partnership through comparative case studies help illuminate different state traditions of governing? Across the different national sites are the very definitions of the problem of partnership. This is illustrated in Silver's (1994–1995) discussion between Anglo-American and French approaches to equity. Anglo-American literature focuses on concepts related to individual access that embody particular liberal constructions of individualism. This liberal construction is embodied in the terms of inclusion as access. French discussions, in contrast, use the term social integration.[2] The term is deployed in relation to a French image of the state as a body whose function is to preserve and further the collective social goals of the nation. The differences in Anglo-American and French words express fundamentally different ways of thinking and action about the relations of the state, the social, and individual. In a similar way, the chapters of this book can be read not only as individual stories about national projects of partnerships but as comparative studies of how particular notions travel and are translated within particular spaces and relations of power.

In considering these questions, the contributors to this volume will critically examine an array of partnership initiatives that include different kinds of linkages, collaborations, and networks as they occur within public school settings and across the spaces of school, home, community, business, church, and state. In addition, the selected contributors will represent these issues in multiple national settings in part to illustrate the commonalities and differences in which the language and practice of reforms around the concepts of partnerships and enhanced democratic participation are occurring. At the same time, the case studies represented by the contributors to the volume will illustrate and analyze differences in the language and practice of reform that depend upon distinct cultural histories as well as variations in the concepts of nation, state, and citizenship.

The significance of this volume to the literature on educational reform is in the critique that each of the contributors will bring to the conception of reform and partnership through their analysis of the language and practices of selected educational initiatives. The different theoretical lenses employed by the volume's contributors will look at the relations between new patterns of globalization, the state, and the new governing patterns that are emerging under the rubric of partnerships, networked societies, and alliances. Contributors will pay particular attention to the politics of partnerships as evidenced in the role they play, both positive and negative, in enhancing the voice and civic engagement of previously marginalized groups. Doing this will involve a consideration of emerging new patterns of governing and new ways to think about power and knowledge both at the level of the state as an institution and at the level of the subjectivities of the ways in which we internalize the notions of opportunity for civic participation, greater empowerment, or partnerships and participation. The idea of community as a global set of partnerships or a network society where the market and democratic participation can act together to promote opportunity and equality will also be examined as part of the new conception of globalization and its implications for inclusion as well as exclusion. Therefore, by posing the problem continually as a rethinking of participation, democracy and the meaning of citizen, the volume, its editors and contributors, will tackle and intertwine the broader changes occurring internationally and within nations with the theoretical and political problems that are embedded in the different reforms embodying notions of partnering.

The essays in this volume are grouped into four sections. The first section explores the interplay between state and changing ways of governing. Thomas Popkewitz' essay (chapter 1) looks at the pattern of reasoning embedded in the discourse of partnerships in contemporary educational reform. R. Buenfil Burgos (chapter 2) looks at twentieth-century Mexican educational reform to explore the ambiguous meanings of the concept of partnerships. The second section considers how partnerships link the state, civil society, and education together. Barry Franklin and Gary McCulloch (chapter 3) compare two partnership ventures, City Technology Colleges and Educational Action Zones. Educational Action Zones are provided extended treatment in chapter 4 by Marney Dickson, Sharon Gewirtz, David Halpin, Sally Power, and Geoff Whitty. Yvette Baber and Kathryn Borman (chapter 5) explore the interplay between schools and communities in two partnerships involving high schools in Miami, Florida.

The essays in the book's third section consider the impact that public–private partnerships have on school reform. Lynn Fendler examines the history of the Michigan Partnership for New Education (chapter 7), which employed partnerships in reforming teacher education. Henry Levin's contribution (chapter 6) considers how partnerships bring together public and private sector initiatives in contemporary American school reform.

The last section of the book considers the problem of inclusion and exclusion as it relates to the governing of partnerships. Miram David (chapter 8) examines how issues of inclusion and exclusion are related to parent–school partnerships in contemporary British educational reform. And Marianne Bloch, I-Fang Lee, and Ruth Peach (chapter 9) examine patterns of inclusion and exclusion in a community–university partnership at the University of Wisconsin–Madison that has been designed to provide tutoring to low-income youth with "potential."

Notes

1. This view of homogeneity is historically questionable as it filters out the projects of constructing the sense of nation-ness among diverse groups and the processes of making a nationhood (see, e.g., Noiriel, 1988/1996).
2. The term social inclusion, however, has been picked up by the Labour Government.

References

Arthur, J. (2000). *Schools and Community: The Communitarian Agenda in Education*. London: Falmer Press.
Berger, R.A. (1985). "Private-Sector Initiatives in the Reagan Era: New Actors Rework an Old Theme." In L. Salamon and M. Lund (Eds.), *The Reagan Presidency and the Governing of America* (pp. 181–211). Washington, D.C.: The Urban Institute Press.
Blair, T. (1997). *New Britain*. Boulder: Westview Press.
Blair, T. (1998). *The Third Way: New Politics for the New Century*. London: The Fabian Society.
Blair, T. (October 1, 2002). "At Our Best When at Our Boldest." Speech by Tony Blair to the Labour Party Conference, Blackpool, England. Retrieved October 19, 2002 from http://www.labour.org.uk/tbconfspeech/.
Bloch, M.N., Holmlund, K., Moqvist, I., and Popkewitz, T.S. (Eds.) (2003). *Governing Children, Families and Education: Restructuring the Welfare State*. New York: Palgrave Press.
Bloch, M.N. and Popkewitz, T.S. (2000). "Constructing the Child, Parent, and Teacher: Discourses on Development." In L.D. Soto, *The Politics of Early Childhood Education* (pp. 7–32). New York: Peter Lang.
Butler, J. (1993). *Bodies that Matter: On the Discourse Limits of "Sex."* New York: Routledge.
Campbell, R.F. and Lorion, J.E. (1972). *Performance Contracting in School Systems*. Columbus: Charles E. Merrill Publishing Company, 1972.
Clinton, B (October 2, 2002). "New Labour and the Third Way Works." Speech by Bill Clinton to Labour Party Conference, Blackpool, England. Retrieved October 19, 2002, from http://www.labour.org.uk/clintonconfspeech/.
Coates, D. (2000). "The Character of New Labour." In D. Coates and P. Lawler (Eds.), *New Labour in Power* (pp. 1–15). Manchester: Manchester University Press.
Comte, Auguste (1827/1975). *Auguste Comte and Positivism: The Essential Writings*. Gertrud Lenzer, Ed., New York: Harper and Row.
Commission on Public Private Partnerships (2001). *Building Better Partnerships: The Final Report of the Commission on Public Private Partnerships*. London: Institute for Public Policy Research.

Cutler, W.W. (2000). *Parents and Schools: The 150-Year Struggle for Control in American Education*. Chicago: University of Chicago Press.

Dussel, I. (in press). "Educational Restructuring in Argentina: Hybridity, Diversity, and Governance after Welfare." In S. Lindblad and T. Popkewitz (Eds.), *Controversies in Educational Restructuring: International Perspectives on Contexts, Consequences and Implications*. Greenwich: Information Age Publishing.

Foucault, M. (1991). "Governmentality." In G. Burchell, C. Gordon and P. Miller (Eds.), *The Foucault Effect: Studies in Governmentality* (pp. 87–104). Chicago: University of Chicago Press.

Franklin, B.M. (1994). *From "Backwardness" to "At-Risk": Childhood Learning Difficulties and the Contradictions of School Reform*. Albany: State University of New York Press.

Fraser, N. (1997). *Justice Interruptus: Critical Reflections on the "Postsocialist" Condition*. New York: Routledge.

Gamarnikow, E. and Green, A. (1999). "The Third Way and Social Capital: Education Action Zones and a New Agenda for Education, Parents and Community." *International Studies in Sociology of Education*, 9, 3–32.

Gewirtz, S. (1999). "Education Action Zones: Emblems of the Third Way?" *Social Policy Review*, 11, 145–165.

Giddens, A. (1990). *The Consequences of Modernity*. Stanford: Stanford University Press.

Giddens, A. (1994). *Beyond Left and Right: The Future of Radical Politics*. Stanford: Stanford University Press.

Giddens, A. (1998). *The Third Way: The Renewal of Social Democracy*. Cambridge: Polity Press.

Gillborn, D. (1998). "Racism, Selection, Poverty and Parents: New Labour, Old Problems?" *Journal of Education Policy*, 13, 717–735.

Gilpin, R. (2000). *The Challenge of Global Capitalism: The World Economy in the 21st Century*. Princeton: Princeton University Press.

Goodwin, R. (1996). "Inclusion and Exclusion." *Archives Europeennes de Sociologie*, 37, 343–371.

Hacking, I. (1999). *The Social Construction of What?* Cambridge: Harvard University Press.

Hatcher, R. and Leblond, D. (2001). "Education Action Zones and Zones d'Education Prioritaires." In S. Riddell and L. Tett (Eds.), *Education, Social Justice and Interagency Working: Joined-up or Fractured Policy?* (pp. 29–57). London: Routledge.

Heffernan, R. (2000). *New Labour and Thatcherism: Political Change in Britain*. London: Macmillan Press Ltd.

Howell, D. (1980). *British Social Democracy: A Study in Development and Democracy*. New York: St. Martins Press.

Kliebard, H.M. (1999). *Schooled to Work: Vocationalism and the American Curriculum, 1876–1946*. New York: Teachers College Press.

Latham, M. (2001). "The Third Way: An Outline." In A. Giddens (Ed.), *The Global Third Way Debate* (pp. 50–73). Cambridge: Polity Press.

Laybourn, K. (2000). *A Century of Labour: A History of the Labour Party 1900–2000*. Phoenix Mill: Sutton Publishing Ltd.

Levitan, S.A. (1969). *The Great Society's Poor Law: A New Approach to Poverty*. Baltimore: The Johns Hopkins University Press.

Levitas, R. (1998). *The Inclusive Society? Social Exclusion and New Labour*. London: Macmillan Press Ltd.

Linder, S.H. and Rosenau, P.V. (2000). "Mapping the Terrain of the Public–Private Partnership." In P.V. Rosenau (Ed.), *Public–Private Policy Partnerships* (pp. 1–18). Cambridge: The M.I.T. Press.

Maurrasse, D.J. (2001). *Beyond the Campus: How Colleges and Universities form Partnerships with their Communities*. New York: Routledge.

Mecklenburger, J. (1972). *Performance Contracting*. Worthington: Charles A. Jones Publishing Company.

Meyer, J., Boli, J., Thomas, G., and Ramirez, F. (1997). "World Society and the Nation-State." *American Journal of Sociology*, 103, 144–181.

Mittelman, J.H. (2000). *The Globalization Syndrome: Transformation and Resistance*. Princeton: Princeton University Press.

Muravchik, J. (2002). *Heaven on Earth: The Rise and Fall of Socialism*. San Francisco: Encounter Books.

Noiriel, G. (1988/1996). *The French Melting Pot: Immigration, Citizenship and National Identity*. G. De Laforcade (Trans.). Minneapolis: University of Minnesota Press.

Nóvoa, A. (2002). "Ways of Thinking about Education in Europe." In N. Nóvoa and M. Lawn (Eds.), *Fabricating Europe: The Formation of an Educational Space*. (pp. 131–156). Dordrecht: Kluwer Publishers.

Popkewitz, T.S. (1976). "Reform as Political Discourse: A Case Study." *School Review*, 84, 43–69.

Popkewitz, T.S. (1998). *Struggling for the Soul: The Politics of Schooling and the Construction of the Teacher*. New York: Teachers College Press.

Popkewitz, T.S. (in press). "Governing the Child and Pedagogicalization of the Parent: A History of the Present." In M. Bloch, K. Holmlund, I. Moqvist, and T.S. Popkewitz (Eds.), *Governing Children, Families and Education: Restructuring the Welfare State* (pp. 35–61). New York: Palgrave Press.

Popkewitz, T.S. and Bloch, M.N. (2001). "Administering Freedom: A History of the Present—Rescuing the Parent to Rescue the Child for Society." In K. Hultquist and G. Dalberg (Eds.), *Governing the Child in the New Millennium* (pp. 85–118). New York: Routledge Falmer.

Popkewitz, T.S. and Lindblad, L. (2000). "Educational Governance and Social Inclusion and Exclusion: Some Conceptual Difficulties and Problematics in Policy and Research." *Discourse*, 21, 5–54.

Power, S. and Gerwitz, S. (2001). "Reading Education Action Zones." *Journal of Education Policy*, 16, 39–51.

Power, S. and Whitty, G. (1999). "New Labour's Education Policy: First, Second or Third Way?" *Journal of Educational Policy*, 14, 535–546.

President Bush's faith-based initiative (n.d.). Retrieved January 21, 2003, from www://whitehouse.gov/government/fbci/fbci_overview.pdf.

Reese, W.J. (2002). *Power and the Promise of School Reform. Grassroots Movements During the Progressive Era*. New York: Teachers College Press (original work published 1986).

Reich, R. (1991). *The Work of Nations.* New York: Alfred A. Knopf.
Rose, N. (1999). *Powers of Freedom: Reframing Political Thought.* Cambridge: Cambridge University Press.
Rose, N. (2000). "Community, Citizenship, and the Third Way." *The American Behavioral Scientist,* 43, 1395–1411.
Schramm, S. (2000). *After Welfare: The Culture of Postindustrial Social Policy.* New York: New York University Press.
Silver, H. (1994–1995). "Social Exclusion and Social Solidarity: Three Paradigms." *International Labour Review,* 133, 531–577.
Silver, H. and Silver, P. (1991). *An Educational War on Poverty: American and British Policy-Making 1960–1980.* Cambridge: Cambridge University Press.
Wagner, P. (1994). *A Sociology of Modernity: Liberty and Discipline.* London: Routledge.

Part I
Partnerships and Changing Patterns of Governing

CHAPTER 1

PARTNERSHIPS, THE SOCIAL PACT, AND
CHANGING SYSTEMS OF REASON IN A
COMPARATIVE PERSPECTIVE[1]

Thomas S. Popkewitz

Policies and research about school reforms embody salvation themes. The modern salvation themes of schooling are not religious in seeking a heaven in the-life-after. They are secular in offering the deliverance of the nation through the education of the child. Salvation themes of rescue, redemption, and progress are embodied in the worldwide institutionalization of schooling as the nation-state formed in the nineteenth century (Meyer et al., 1997). Contemporary school reforms are spoken about as insuring the future of democracy in the new world of, to use its planetspeak, "a global" and "knowledge-based" society.[2] Partnerships in educational reforms are one such salvation theme. From different ideological positions, partnerships tell of collective progress through promoting civic participation through individual and group involvement in the local agencies. The stories of educational partnerships are tales of seeking a newly arrived consensus and harmony between the governed and the government.[3]

The salvation themes of educational partnerships do not stand alone as a force of progress and redemption. They embody a particular double relationship of governing. At one end, *partnerships* tell of the state's collective obligation to promote justice and equity through the schools. Partnerships are mandated practices of educational reforms assessed through state fiscal and educational outcome accountability procedures. And partnerships also narrate the responsibility of the community and individual to participate as agents of progress. This double relationship is narrated in the American story of political pluralism. That story is about the state's obligation to promote the common good, such as symbolized in the need for a minimum wage, child labor laws, and universal education. I call this side of governing as the Pact, the inscription of collective norms in state actions of reform. The "other" side of governing is the formation of civil society, the partnerships where different associations, groups, and individuals engage in the self-governing responsibilities and obligations of citizenship. This "partnership" side of government is expressed as schools, communities, and business groups working together to produce both individual agency and the general development of society.

The relation of the pact and partnership are significant as they produce governing patterns. What seems as two extreme poles of collective norms and individual involvement are in fact mutually related in producing concrete sets of principles of action and participation. There is no state without a civil society; and there is no self-governing without the conditions of the state that produce the calculus of governing action and participation. The salvation themes of educational partnership, then, embody governing practices formed through this mutual relationship.

This chapter examines the relation of government and governing embodied in educational partnerships. The governing is related to the cultural practices of reform that produce how judgments are made, conclusions drawn, rectification proposed, and the fields of existence made manageable and predictable in school reform (see, e.g., Popkewitz, 1998b; Popkewitz and Lindblad, 2000). While we rarely think of reform as cultural and governing practices, the narratives and images of the child and teacher in school reforms embody ways of living, "seeing," thinking, and feeling. To think of governing in this manner is to recognize that the cultural practices of educational reforms are political practices that generate principles about action and participation.

This study of partnerships as a governing practice draws from a multi-European-Union-country study of reform policy and U.S. school reforms.[4] The first section considers two seemingly opposite sides of a political theory of participation and responsibility in education. One side is the political pact. Educational partnerships exist within state-centralized policies for school management and system accountability that are made in the name of the collective obligation. That obligation is, for example, to ensure "quality control" and equity through achievement (outcome) testing or other evaluative schemes in efforts to change the school curriculum and teaching. The other side of political theory are decentralized policies to accord "communities" with a greater role in decision-making and to promote a self-governing individual. One of the foundations of partnerships is the call for self-managing individuals and groups through decentralization of school finances and decision-making. But centralization and decentralization are not separate practices. I argue that they embody patterns of governing that relate collective rationalities and the self-governing of the acting subject.

The second section pursues the governing as inscribing "ways of living." It examines the field of cultural practices in pedagogical reform. That is, partnerships are not only about institutional changes but embody cultural practices of who the child is and should be. These practices are to "make" a self-reflective lifelong learner, collaboration as a measure of a teacher's and child's "empowerment," and the alchemy of school subjects. The alchemy of school subjects gives attention to the representations of an expert world that fashion the curriculum rules and standards for the child to act and participate. The third section explores the relation of the pact and partnerships as a simultaneous system that excludes as it includes. The normalizations and divisions of pedagogy produce principles about who the child, parent, and teacher are, should be, and the characteristics of who does not "fit" those norms of participation. The final section relates the analysis of the governing patterns of schooling to other social and cultural fields in forming principles of action and participation. Briefly examined are changes in the fields of economy, philosophy/social science, and military as they relate to education. I argue that there are particular boundaries for human agency that are homologous to the cultural practices of education. I use the word homology to explore how the different cultural practices overlap in different fields in ways that are not in correspondence to each other or in a causal relation.

My interest in partnership, then, is not whether it is good or bad, but as a site to explore a significant "fact" of modernity: power is exercised less through brute force and more through the systems of knowledge or reason in which the objects of schooling are made comprehensible and capable for action. The educational reforms that call for partnerships exist within a field of cultural practices that social theories of education need to interrogate. The knowledge of the child, teacher, parent, and community in school reform are not "merely" descriptions or a representations of human intent but political. Political in the sense that inscribed is a continuum of values about the objects of reflection and action, and who is capable of that participation.

The Pact and Partnership: Reconstituting the Governing of the Schooling

Reforms about educational partnerships are offered as salvation themes about democracy and participation in the system of schooling and organizational changes of schooling occurring across Europe and North

America. The partnerships exist, ironically, in relation to centralizing practices of school management designed to increase school efficacy and efficiency. Policy gives attention to the management of the school. New management schemes are installed to signify the collective obligation of the state for ensuring equity and social progress. I call this notion of the state as representative of the collective interest of the citizen as the pact of government to its citizens. The pact is formed symbolically as a social contract or agreement between the government and its citizenry to ensure progress and individual well-being. But traveling with the centralizing practices of management are practices of partnership that decentralize through creating processes of community involvement. The decentralization strategies about school–community partnerships, for example, are to re-create the political contract. While educational research divides the pact and partnership, they overlap in practice as individual reflection, interpretation and identity are also objects of social administration.

Re-Visioning the Principles of Governing in the Public Management of Schools

To consider the relationship of the notions of partnership and pact—of the workings of civil society associations and networks to collective obligations—I first focus on the new public management schemes for school accountability. The introduction of new public management and financial controls is offered as a way for the state to rethink the pact by making the school more clearly accountable to social goals, especially in times of decentralizing many fiscal, personnel, and curriculum controls of previously centralized, state-organized controls of school systems. Thus, the public management and new financial controls are deployed within a field of practices that decentralize school policy and the school actor. The new public management approaches not only renew the pact but they also embody new sets of relations of a centralized and decentralized system that, I argue, vests individuals with particular capacities and capabilities.

Diverse places as Britain, Iceland, and New Zealand (Marshall, 2000; Alexiadou, Lawn, and Ozga, 2001; Ingólfur Ásgeir Jóhannesson, with others, 2002) have instituted accountability measures as a salvation narrative. The salvation theme of the reforms is that proper measurement and assessment of schools will ensure that all citizens have adequate opportunities in a society that strives for social justice and equity. Icelandic school reforms, for example, embody a narrative about progress based on efficient school management. A saga is told of more efficient, evaluative, and regulative practices for monitoring classroom learning and school resources that will ensure equal outcomes. National policies of contract management approaches and performance objectives in budget reforms are to make the Icelandic schools the best in the world. One element of this notion of performance is schools receiving money in proportion with the number of students and credits they finish. This notion of "best," however, exists with other practices about school change and reform. For example, the Icelandic government uses the international indicators of science and mathematics achievement (TIMSS) as a device to argue both for the new management controls and as a guide in rectifying the Icelandic curriculum. The assessment data of TIMSS, curriculum changes, and performance-based education are placed in proximity to each other as redemptive narratives about the state's commitment to social and economic progress.

New central controls in Britain maintain a similar saga about salvation. The management schemes of Teaching Quality Assurance (TQA) are instituted as necessary for the state to perform its obligations in education. The TQA is a management scheme. The English National Curriculum and its assessment packages, for example, are thought to provide ways of standardizing schooling so parents can make informed choices, teachers can be trained to deliver those standards, and inspectors can provide more objective and consistent judgments of school improvement (Alexiadou, Lawn, and Ozga, 2001). The management schemes involve the publishing of school performance data. Checklists of skills to be taught to children are identified and measured individually and used to measure school improvement. The skills are thought of as those that are to be transferable to employment (see, Smith, 2000). The improvement of schooling is to calculate and divide the "teacher" into molecular skills that are evidence of adequate performance. The TQA is used to manage the training of teachers through identifying measures of their performance. In both instances, the TQA is formulated as a measuring device to enable the state to guarantee its responsibility to its citizens (the pact) through making visible the schools' performance to all citizens. The reforms move into the university as performance objectives that articulate a relationship between professional certification criteria and national commitments for a more progressive and globally competitive society.

The reforms that are to signify the *pact* are located with other reforms that are to decentralize decision-making in a reconstruction of the state with civil society. The political language of this new relationship is often phrased as a *partnership*. British TQA calls for partnerships in Educational Action Zones. The zones are collaborative relationships of businesses, local governments, local parent groups, and the state (Alexiadou, Lawn, and Ozga, 2001). The centralized TQA and the decentralized partnerships overlap in a salvation narrative about the needs of the new economy and social progress toward a more equitable and just society. One can find a similar rearticulation of the pact with the partnership, between central and local decision-making, in U.S. educational reforms that bring together national professional teaching and curriculum standards with local partnership initiatives such as charter schools, voucher systems, and "choice" programs.

The relation of the pact and partnership in the reforms has a double sense of governing that is captured in the phrase, "We, the People." The "We," in the "We, the People" represents the commitments of the pact found in the policies of the TQA. The TQA is to represent the collective obligation of the state to maintain a common, collective good through the school. But locating the TQA with partnerships evokes another part of the "We" in "We, the People." That other "We" is located in the individual, groups, and local community participation from which the democratic social contract is to be forged from the bottom-up.

The joining of the pact and partnership, however, is more than organizational changes in who make decisions. The amalgamation of practices enacts procedures for ordering how judgments are made and the field of existence for the educational actor (Popkewitz, 1996). Let me pursue this through an example of Swedish reforms. After World War II, the consolidation of state responsibility for organizing and evaluating school subjects became more pronounced. The centralized management of society by the Swedish welfare state was to ensure the collective social and economic progress. The school was a rule-governed system defined through parliamentary legislation and the strong, centralized bureaucracy of the Swedish Board of Education. The state was to secure equity and justice through equal opportunities in educational access. An elaborate system of statistical studies of school progress and social class was organized to evaluate and guide policy.

The new monitoring agency of the Swedish Agency for Education (*Skolverket*) was created in 1991 by the Social Democrats. It was charted to monitor a decentralized school system. The reform program, however, involves a centralized goal-driven conception of the state vis-à-vis the educational arena. The new pact was expressed as the national setting of school curriculum goals, including a more flexible time allocation in school subjects. The school goals were "steering" mechanisms to enable a plurality of local community solutions in finding the paths toward those goals.

The Swedish reforms involve new capabilities and capacities of the systemic actors in education. Previous state rule–governed patterns imbued a feeling of certainty and control in Swedish professional practices. Now, state officials monitor school practices with an uncertainty and flexibility as there are no fixed rules of application. The reconstructed official encounters multiple, locally grown solutions rather than prescribed answers to educational problems. In the language of the new reforms, flexible problem-solving empowers the new school actors who "learn in cooperative groups" and "construct their own meanings."

The centralized Swedish goals also entail decentralized assessment systems to monitor curriculum goals. National survey and testing approaches work alongside "qualitative" research that makes legible the norms and local culture of classrooms in the new decentralized reforms system. Teachers are to practice their own self-monitoring and evaluation of children's careers in school. Teacher education, which previously had little systematic coursework in teacher organized evaluation, now has new courses of study. Portfolio assessment and action research in the training of teachers are introduced to focus on the new local responsibilities of the teacher to evaluate in the classroom.

Similar movements that center and decenter school governance are expressed in the U.S. notion of "systemic school reform." The state coordinates governmental agencies, professional teacher groups, research communities, and local authorities to provide a flexible but coherent policy of reform (see, e.g., Smith and O'Day, 1990). Charter schools, school choice, and teacher professionalization strategies coexist with national standards and testing in the United States. The centralized and decentralized reforms embody the system of reason found in Sweden but often with different ideological statements.

Thinking that the organizational and institutional changes produce the changes in the governing patterns is easy. But this sense of causation would, I think, misplace the complex patterns of governing. The changes involve an overlay of different cultural practices embodied in state planning, research programs about teaching, and the organizational configurations of schooling. Hultqvist (1998) argues, for example, that the Swedish education reforms were made possible through the decentralized, child-centered pedagogy of the Swedish teacher introduced in the 1970s. The prior discursive practices make possible the governing principles that shape and fashion the kinds of individuals who are productive in the organizational patterns of centralization/decentralization.

Community and Governing Principles of the Pact/Partnership

A different side of the notion of partnership is expressed in the notion of community that travels along with educational policy and research programs. Community is a seductive metaphor in today's political landscape of reform. Current Finnish, Swedish, U.S., and British reforms use the metaphor of community to articulate both the notion of civic involvement and political responsibility to the moral and ethical commitments of democracy. Community emphasizes the idea of the autonomous citizen (the partnership) from which the collective contract and moral good are constituted. The metaphor of community is evoked in the different U.S. reforms of site-based management, home–school collaboration, and curriculum discussions of the classroom as a "community of learning." Community provides a metaphor to direct attention to inclusive programs as it embraces diversity, difference, and multiculturalism. It is also evoked in discussions of school choice to narrate how those communities where parents have no choice in schools can now choose.

The notion of *community* provides a way to consider the double relationship of the "We, the People" in the pact and partnership. The civic ideal of community is related to the sublime in which the home of God is found (Cronon, 1996). In the early nineteenth century, the landscapes of nature were the places where one has a better chance than elsewhere to glimpse the face of God. This idea of nature is itself expressed in the idea of the New World, initially a religious phrase that evoked images of the Garden of Eden. Community became the new cathedral by which individuals could be brought into a face-to-face relation with God's creations of Nature. In much of the urban planning that took place in the nineteenth century was the installation of ordered gardens and parks in which communities can better function in relation to God. But the notion of community was also an expression about the return to the conditions in which neighbors form the democratic institutions that predate modernity.[5]

The Protestant religious motifs about the individual's good work and liberal democratic commitments through the notion of community were inscribed in the formation of the American social sciences. The Chicago School of Sociology interactions with the Hull House settlement program, for example, articulated the liberal and religious themes of community in reforming the urban setting. Diverse ethnic communities of immigrants were to be joined into a more general social and collective identity related to nation building. The reinstalling of face-to-face interactions through theories of the family and community were to produce democracy and its images of the sublime. The writings of John Dewey who worked with the Chicago reformers and social scientists deployed the notion of community to develop a collective, American identity that reworked the relation of religious and political images in the notion of the child. Community is a word resurrected in today's reform discourses to express different images of the child expressed in twin terms as empowerment and social solidarity.

The double relation of the "We, the People" is ironically evident in the reorganization of the welfare state and its emphasis on community. In an influential book about the Third Way, Anthony Giddens, a British sociologist and advisor to the Labour government of Tony Blair, captures social and educational alignments that become apparent in the 1990s.[6] The Third Way articulates the pact as the state's forging partnerships for community renewal and development. For Giddens, government sheds the old ideological divisions of conservative, neoliberal minimalist policies, and social democratic social welfare intervention traditions to reinscribe a government that is at once centralized and decentralized. The Third Way, Giddens argues, is a "movement of double democratization" as the state organizes the expansion of the public sphere and plays a major part in "renewing civil culture" through fostering partnerships that involves a "redistribution of possibilities" (Giddens, 1998, p. 109).

Giddens places the relationship of the pact and partnership as central to the progress of society and the moral commitments of the state. He argues, "there are no permanent boundaries between government and civil society" (Giddens, 1998, pp. 79–80). Governing is where "the contract between individual and government shifts, since autonomy and the development of the self—the medium of expanding individual responsibility—becomes the prime focus" (Giddens, 1998, p. 128). The greater harmonization brought by educational standards and public partnerships create a lifelong learning that, Giddens argues, is the friend rather than an enemy of diversity.

The doubleness of the pact and partnership is eluded in the mythology of the modern American school. Current policies, research, and histories celebrate the school governed by local decision-making and through regional authorities with no centralized bureaucracy. The mythology of American schooling is told as variations of a seemingly stateless state of partnerships in which various local and national coalitions work toward reform. Ravitch (1996), for example, discusses the development of national curriculum standards and testing as a process of reform that is "setting a new course in a democracy (xvi)." The creation of national standards, Ravitch argues, needs to be considered in a tradition of local, control and the negotiations of various state, local, and professional organizations. Ohanian (1999), rejecting national standards as an affront to democracy, installs a version of democracy that calls upon the local teacher, parent, and community as knowing what is best for children. The seemingly oppositional interventions of Ravitch and Ohanian are iterations of salvation themes of *democracy* that evoke the myth of local, collaborative action to revive the republic (Popkewitz, in press).

What these accounts obscure is the system of reason that historically circulates to order the interpretation, reflection, and action. While many countries are engaging in reforms that move in the opposite direction of U.S. reforms through decentralization practices, the relation of the local and the national in governing still needs to be considered. There is no local collaboration without a system of classification that is constituted as "preexisting facts" and principles that shape and fashion what is possible as decision-making (see, e.g., Schram and Neisser, 1997; also, Popkewitz, 1998a,b). The centralization and decentralization practices are involved in a field of cultural practices whose rules and standards of reason substantively alter social relations and individual and collective responsibilities.

The New Democracy: Salvation Themes of Science in the Production of the Collaborative Community

At this point, I move to the particular distinctions and differentiations about the child and teacher in reforms about partnerships. My argument is that the reforms that call for partnerships and accountability are located in a field of practices that intersect with pedagogical principles that generate principles about the kinds of individuals that children are and are to become. The concrete practices of pedagogical discourses are embodied in an amalgamation of cultural practices that link individuality to collective senses of obligation and salvation, what I earlier discussed as the relation between the pact and partnership. When policy, research, and pedagogical texts are read in proximity to each other, particular ways of thinking and ordering the objects of action can be made visible. The pedagogical reforms place notions of collaboration and partnerships in a field of practices related to the action of the child who participates, the professional teacher who works with parents and communities, and a pedagogical content of school subjects to organize what parents and teachers talk about when being in a partnership. In the amalgamation of practices is the relation of the pact/partnerships through interstices in which principles are produced to govern the child and teacher who are to act in the reformed school.

The Self-Reflective, Problem-Solving Lifelong Learner
Policies about the new child, as the system reforms discussed earlier, are made in the name of democracy and liberty in a global world. The salvation theme is of future economic progress and the promise of equity and justice in schools. Yet the globalized child is continually placed within the narratives of the nation—the obligation of preserving the democracy of America or the social democracy of Sweden or Finland in a global society. The teacher in the U.S. reforms, for example, is one who enters partnerships. These partnerships are to create a new leadership that is "energized" to "work with others" to "ensure that America and its children will have the schools they require and deserve" (p. ii), and to provide "a down payment to renewal and reform" that the "American public" demand so (p. 1) "the nation's schools can and must serve better the citizens of our democracy . . ." (American Council on Education, 1999, p. 1).

The double of the "We, the People" that constitutes the pact and partnership is embodied in the national standards for teachers. The salvation narrative is one of proper administration of democracy and provision of an inclusionary society (see, Darling-Hammond, 1998). The national standards, however, do not stand alone but within a field of reform strategies related to new pedagogies. Teacher education reforms, the Carnegie Council on Adolescent Development (1995), the American Council of Education (1999), and the National Council for Teachers of Mathematics (2000) curriculum standards, among others, articulate salvation themes about remaking the child who is to act responsibly in the future as an American citizen in the global world.

What is the responsible child of the future that is inscribed in the practices of pedagogy? The redemptive language of modern pedagogy is to help children become good citizens, better adjusted people, and active learners. The new child in pedagogy is homologous to the capabilities of the Swedish State official. Both the civil servant and the child embody characteristics of working flexibly as a problem-solver in uncertain contexts that have no fixed rules of application or prescribed answers to educational problems. But the child does more than merely solve problems. In the language of pedagogical reforms, the child constructs knowledge in communities. A cornerstone of the reforms is a social–psychological and psychological constructivism. Learning mathematics or science, for example, is developing flexible approaches and responding to new eventualities as there is no longer only one correct answer. The child, to use the language of contemporary reforms, is an active, self-reflective problem-solver and lifelong learner who operates with no single center and fluid boundaries (Popkewitz, 1991).

The notion of a lifelong learner involves a calculus of intervention and displacement of the ethical obligation for the child. The reforms are told as sagas of democracy, modernization, and the globalization of the nation. But this version of democracy seems to have no collective social identity except in the collection of different communities themselves (Alexiadou, Lawn, and Ozga, 2001; Lindblad, Lundahl, and Zackari, 2001; Simola, Rinne, and Kivirauma, 2001). The empowerment is in the individual who constructs and reconstructs one's own "practice," participates in collaborative problem-solving, and self-manages the autonomous ethical conduct of life (see, e.g., Rose, 1999; Rose and Miller, 1992). The self-management of one's personal ethics is in opposition to the early part of the twentieth century where teachers were to "educate" the child to universal rules that linked the individual to collective national sagas. The child's participation in communities was to externally validate social morals and obligations of the citizen of the nation.

The site of intervention is the soul. Modern pedagogy is to shape the inner capacities and capabilities of the child. While many might object to the interjection of the idea of the soul in discussions of the governing of schooling, the soul is a regulating impulse in the formation of the early republic that merged religious notions of salvation and redemption with the governing of the state (see, e.g., Ferguson, 1997; and more generally, Foucault, 1979; Rose, 1989). The soul of the modern school is of an individuality ordered through the dispositions, sensitivities, and awarenesses that "make" the civilized actor who he is today to operate in a global culture and economy.

The New Expertise of the Teacher: A Partner in Pact and Partnership
The policy and research programs of teacher education embody a new type of expertise that needs to be placed within the field of cultural practices discussed earlier. The teacher-as-expert is one who engages individuals and communities in partnerships so that they can be better managed, healthier, and happier. Learning is related to an expertise of the teacher that is not to assess the truth of statements but is to govern the dispositions and sensitivities of the child. The reformed teacher is to coach and to facilitate. As with the child who is a lifelong learner, the teacher is also signified as a lifelong learner in many policy statements.

The language of teacher education reform is to develop a "reflective" teacher. That teacher is one who is responsible for "problem-solving" in a world that is personally unstable. The professional teacher is "self-governing," and has greater local responsibility in implementing curriculum decisions for children's learning. The capabilities and capacities of the teacher and child are homologous but not reducible to the sensitivities and awarenesses inscribed in the pedagogical constructivism and the decentralized and centralized organizations discussed earlier.

The new expertise of the teacher, as I have argued earlier, involves a field of cultural practices. The amalgamation of such practices, for example, is embedded in the American Council on Education (1999) reforms on teacher education. In this statement of reform, it is possible to see different cultural practices that link universities in partnership with other institutions in generating principles about what the child and teacher are. Reform, it asserts, involves "(a) a common vision of good teaching; (b) well-defined standards of practice and performance that guide and measure courses and clinical work; (c) a rigorous core curriculum; (d) extensive use of problem-based methods, including case studies, research on teaching issues, performance assessments, and portfolio evaluation; and (e) strong relationships with reform-minded local schools" (American Council on Education, 1999, p. 5).

In this quotation, a number of practices overlap to order the principles of action and participation. The notions of collaboration, the rules that constitute "problem-solving" of the child and the teacher, and the notions of community and parent participation are placed together to classify the "experience" of the teacher. The self-actualized teacher is one that remakes her biography through continually calculating and rationally researching one's self. The new pedagogical narratives of the teacher map, classify, and work on the territories that constitute the individuality of the teacher and the child. New assessment methods of *performance assessments and portfolio evaluation* function to simultaneously enable the teacher's self-supervision and public observation. *The reflective teacher* of teacher education reforms, as Zeichner (1996) argues, may simultaneously create the possibility of increasing teachers' involvement while also isolating and creating illusions of democratization.

The Alchemy of the Curriculum
The reforms that call for new management and new partnerships take for granted school subjects. The reforms are to make the systems that represent physics or history more effective teaching practices. But school subjects can also be understood as elements in the field of cultural

practices that construct the characteristics of partnerships in the governing of schooling. As the sorcerer of the Middle Ages sought to turn lead into gold, modern curricula produce a magical change as they are transported from the social spaces of historians or physicists into the social spaces of the school (Popkewitz, 1998b). The alchemy transports the disciplines of physics, history, or literary criticism into a psychology of the child. Learning physics is about "concept mastery," the psychology of "cooperative small group learning," and the "motivation" and the "self-esteem" of children. School subjects are performed in relation to the expectations of the school timetable, conceptions of childhood, and conventions of teaching that transform disciplinary knowledge and inquiry into strategies for governing the *child* who is an *active, problem-solving learner*. The only thing left of disciplinary practices when they arrive in school is the namesake—physics or history.

The curricula of mathematics, physics, literacy, and history are not about the cultural practices in which academic knowledge is produced but is about the ordering of the capabilities and dispositions of the child. The alchemy provides a particular translation vehicle that stabilizes academic knowledge in order to make the child as the site of administration. The fixing, stabilizing language is revealed in the curriculum. School subjects are classified as "bodies of knowledge" or "content knowledge" (concepts, generalizations, and procedures) that children learn.

Once fixed, teaching content knowledge focuses on a pedagogy that calculates changing children's capabilities and capacities. Mathematics education is an example. There is talk, for example, about children participating and collaborating in learning mathematics, but the major concepts of mathematics are those of children's development stages and theories of learning and cognition—not of mathematics. What is portrayed as the reason of mathematics is not mathematical reason but the transportation of a psychology into a pedagogy directed toward the inner capabilities of the child.

Mathematical reason, for example, is defined as "the development and justification of use of mathematical generalizations" (Russell, 1999) or examining the "innate reasoning" of the child (see, e.g., Malloy, 1999). Evidence of student–teachers' understanding is measured, for example, by their *conceptual understandings or cooperative learning*. Learning is defined as finding multiple ways that make apparent the presupposed logical and analytical foundations of scientific propositions or mathematics properties. The logic of the academic disciplines is controlled by the psychology of children's development that transmogrifies the complex social and cultural practices that embody mathematics

(Van Bendegem, 1996). Even when there is mention of the complex world of mathematical thinking and its "community of discourse," the words are more of a homage to the differences between the academic field and curriculum than to any systematic attempt to think about the images and narratives of teaching.

The transportation systems of academic disciplines into the categories and distinctions of pedagogical systems should not be surprising. At one level children are not scientists or mathematicians. What is surprising is the particular alchemy. The psychologies of teaching school subjects were not invented to consider the production of academic knowledge, but to administer the conduct of the child. The psychologies of childhood, learning, and cognition in curriculum are inventions that have purposes different from those of understanding and translating disciplinary knowledge into pedagogical problems (Popkewitz, 1998a). For instance, Dewey's scholarship on participation and community embodied cosmopolitan values that were to challenge various processes of modernization in the early twentieth century. Vygotsky's psychology brought the ideals of Marxism into the upbringing practices of the child. And G. Stanley Hall combined romantic visions, Christian ethics, social biology, and science into notions of growth and development.

The alchemy of school subjects is embedded in the amalgamation of cultural practices discussed previously to construct the objects of interpretation and what constitutes the evidence and experiences of teaching. A calculus of intervention and displacement is constituted by the principles generated about the kinds of people that children and teachers are and are to become. But the cultural practices are more. The capacities of autonomy, participation, and collaboration are neither neutral practices nor merely normative commitments. They embody enclosures and internments.

One enclosure related to the increased participation and collaboration among students is evident in the new science textbooks. An examination of science textbooks (McEneaney, 2003) suggests a dramatic pedagogical move to greater student participation, greater personal relevance, and emotional accessibility in science textbooks. The changes, though, insert the iconic image of the scientific "expert" with a particular authority through wider claims of the natural world as ordered and manageable through science. Thus, while there is emphasis on participation and collaboration in classrooms, that cultural space of participation is less and less as the expertise of science makes wider claims about the natural world. These claims about the expertise of science inscribe a certain crystallization and naturalness that are not amenable to human intervention.

Educational Knowledge as the Production of Inclusion/Exclusion[7]

The construction of governing that relates the pact and partnerships, I argued, entails distinctions about the child who embodies the characteristics of problem-solving, development, or learning to make conjectures and justifications. These distinctions and differentiations operate as a map that "says" what dispositions and sensitivities are valued, and what capacities and characteristics of the child and teacher make the characteristics of the "reasonable individual." But as with all maps, the rules and standards of interpretation also impose its opposite. A continuum of values is constructed that differentiates and divides in an unequal playing field. The mapping of the problem-solving child simultaneously embodies distinctions about the child who is not capable of these actions placing that "other child" outside the mapping of "reason" and thus deviant.

This opposition is signified as the child who "lacks self-esteem," the "inner-city child who needs special remediation," and the child "at-risk." The distinctions inscribe theories of deviance that travel as unspoken values about normality. It is in the intersection of normality and the other that exclusion exists. Exclusion is not something different from inclusion but is continually part of the background of something included. In this section, I point to the cultural practices embodied in the reforms that call for partnerships while inscribing a continuum of values that normalize, divides, and exclude even as policy and research seek greater inclusion.

The normalizing and dividing can be illustrated in a study of urban and rural education (Popkewitz, 1998b; also see Mirón, 1996). I use the notion of American urban education as a way to consider a particular response to the inequities and differentiations produced in the practices of schooling. Urban education is one of the salvation stories of U.S. efforts toward an inclusive society. It tells of the state's commitments to correct the wrongs of poverty and discrimination in working toward an equitable and just society. While the notion of urban education is not used in many nations as the poor cannot afford to live in the central city (e.g., Paris, Stockholm, and Buenos Aires), my focus on urban education is an exemplar of a broader set of cultural practices that classify particular populations of schoolchildren in need of "rescue" from poverty and social disintegration.

In the policies and classrooms studied here, the notions of *urban* and *rural* appear as geographical concepts. They are not! They are cultural ones that inscribe distinctions and divisions about normalcy. If I take the notion of urban, it has different meanings that depend upon how the term is placed. There are children who live in the high-rise apartments and brownstone homes of the city, for example, who are not classified in the space that is occupied by the child in programs of urban education. If I can play with the word, these children of the brownstone appear as urbane and cosmopolitan, and not "urban." At the same time, the urban child is someone who lives in cities, the suburbs, and sometimes even in places considered geographically as rural.

What brings the different children together are particular classificatory and normalizing practices that establish difference and deviance. The urban-ness of the child is a theory of difference and deviance. This is not the intention of policy or researcher but of the field of cultural practices in which the urban-ness of the child is fabricated. Urban education is to rescue the child from low self-esteem and a dysfunctional family; and single parent households of low income that have high rates of juvenile delinquency and are without books to read, and so on. The urban child is categorized as the school leaver, disadvantaged, at-risk, and in need of rescue through remediation. Urban education embodies discourses about the disadvantaged, the needy, and at-risk child. The ongoing practices of research and teaching invert the negative norms of the urban-ness into positive elements of instruction (Popkewitz, 1998b). The urban child is signified as the child who learns through "doing" rather than through abstract knowledge and thus has different learning styles from "other" children. Teachers perform with different teaching styles in order to address the differences in the capabilities of the child. The transformation of the negative characteristics into positives ones makes it not possible for the urban child to ever be "of the average."

Territories of membership are produced between members and nonmembers. But the members are not set in the categories of population groups, per se. The sets of distinctions and divisions about the urban teacher and child (and parents) are about the capabilities and capacities that fit and do not fit as problem-solving, reflective, and lifelong learning.[8] While the purpose of public policy and research on urban education is related to principles of equity, the practical consequence is to place the child's *being* as outside of normalcy.

While counterintuitive, the children of urban and rural education are placed in the same classificatory systems of an "urban-ness," at least in how judgments are made about their capacities and capabilities in schooling.[9] In interviews and observations, there were no discursive distinctions between the urban and rural teacher and child. The same sets of categories and distinctions about the child ordered and normalized the child who did not learn and who needed remediation. The child in "urban and rural education" embodies unspoken images and narratives that stand in opposition to the capabilities of the flexible and problem-solving urbane child.

The distinctions and differentiations that make for the "other" in schooling can be related to the work of the sociologist Pierre Bourdieu. Bourdieu's (1979, 1984) study has enabled us to think of the production of differences through the differential systems of recognition and distinctions that divide and organize people's participation. For example, Bourdieu examined the systems of recognition and distinctions among French primary teachers, secondary teachers, professionals, and engineers in how they "appreciated" art, organized their homes with furniture and art, as well as made choices about food, movies, and education. These patterns of distinctions and appreciations were different from, for example, office workers and small shop salespeople. Bourdieu (1989/1996) also explores how the school system consecrates a social nobility through performing a series of cognitive and evaluative operations that realize social divisions. To use Bourdieu's study here, partnerships overlap with distinctions about the child who problem-solves and the urban children to form an unequal playing field built on different characteristics and capabilities of the individual.

Changing Social Fields and School Reforms

The notions of participation, collaboration, and partnership that I discussed here are not only inventions of the system of reason in school policy and research, they are homologous to other institutional settings. The relations are a Wittgensteinian sense of a family resemblance. In this section, I briefly focus on the homologies in the changes in education with those of academic disciplines, the military, and the economy. My references in this section are primarily to those of the United States but also to European literature in order to provide a comparative perspective between educational changes and those of broader social arenas.[10] In these different social spaces are realignments of the relation of the autonomous individual (the partnership) and the constitution of the collective contract of the moral good (the pact).

The notion of partnership/pact in current school policy can be related to a constructive psychology in which children are viewed as "constructing their own knowledge" and in which cooperation and collaboration are seen as essential elements of a child's learning. But constructivism is not only about the psychology of the child. It also relates to debates about "constructivism" and the social construction of knowledge circulating in discussions in philosophy, the social sciences, psychology, and the sciences (for general studies, see Hacking, 1999; Latour, 1999; in education, see Cherryholmes, 1999; Popkewitz, 1997).[11] Since at least the end of World War II, a "constructivist" epistemology is a phenomenon whose elements are found in the rethinking of pragmatism in the social and educational sciences (see Cherryholmes, 1999; Stone, 1999); feminist studies of the "making" of woman-ness (Butler, 1993); studies of social science (Danziger, 1990; Clifford, 1997) and the philosophy of education (Kohli, 1995); as well as critical cultural studies of education (see, e.g., McLaren and Giarelli, 1995; Popkewitz and Brennan, 1998) and postcolonial studies (see, e.g., Chatterjee, 1993; Bhabha, 1994). Constructivism travels in different historical trajectories among intellectual communities that range from Russia, the Scandinavian countries, and the United States, among others. The intellectual discussions in philosophy and the social sciences tend to be more historical, cultural, and social than in the constructivist psychologies found in education. The latter tends to consider the mind in isolation of context or specifically limited to interactional conversations in which norms and actor positions are socially negotiated as found in the idea of a situated cognition. The latter psychologies connect the intimate relations of the individual with the public spaces and the collective norms such as embodied in notions of collaboration.

The constructivist, problem-solving individual discussed earlier in education is also found in the workplace (see, e.g., Gee, Hull, and Lankshear, 1996). The new worker and the new work environment are guided by the "the law of the microcosm." The new work context is flexible through a horizontal structure that enables the development of specific projects that do not have rigid management hierarchies (Fatis, 1992). The smaller work units are said to "empower" workers and to develop flexible, responsive environments in which workers can respond quickly to customer demands. But the pragmatic individuality is also the site of an individual reworking of one's self capabilities and potentialities: "Instead of defining the individual by the work he is assigned to, [we] now regard productive activity as the site of deployment of the person's personal skills" (Donzelot, 1991, p. 252). The work context of technologies (e.g., robots), organization principles (such as "just-in-time" production), and new materials have, it is argued, re-visioned the production process and the worker (see, e.g., International Labor Organization, 1994). The principles of participation of the new economy, it is important to recognize, are not equality distributed within nations, and comparatively across nations, new fault lines of wealth and poverty are produced.

Finally, the notion of the warrior in the military also embodies new objects of interpretation, action, and identity. The military might seem as an odd site to visit in relation to the school, but it is not. The disciplining technologies of the military have been important in the construction of schools since the eighteenth century (see, e.g., Dussel, Birgin, and Tiramonti, 2000). One cannot account for the school testing and measurement industries without paying attention to the problems of recruitment and organization of soldiers during the two world wars (Herman, 1995); nor the emergence of the cognitive science without the defense department's sponsorship of artificial intelligence research for its training and system management (see Noble, 1991).

But what is important to this discussion are the shifts in the governing principles of the subjectivities of the soldier during the past decades. Prior to World War II, a hierarchy of decision-making organized a chain of command in most armies. Furthermore, the technologies of the soldier involved a relationship with the mechanical hardware of war. The airplane, tanks, and guns were practices mastered through theories of mechanics and physics. The discipline of the fighting person involved learning competence in maintaining and applying the mechanical technologies in a hierarchy of command.

Recent rapid changes in military technologies have required different types of performances and "systems of communication" embodied in the soldier. The speed in which decisions are made and the contingencies of the battle make for a world of instabilities and pluralities. There is a need for pragmatic actions as individuals interact with dynamic communication systems rather than fixed mechanical systems. Command structures are changed where today there are both vertical and horizontal axes, or if we keep the language of earlier discussions, strategies of a centralization/decentralization system. A centralized command is becoming more standardized through weapons procurement and technologies that can control battle information in Afghanistan from Tampa, for example. But at the same time, there is a decentralization as the military control structure makes professional flexibility and responsibility a precept of battle discipline.

The different arenas of intellectual thought, economics, and the military overlap and provide a historical specificity to the current discourses about the professional teacher and the child who is self-confident, self-disciplined, and has a capability and willingness to learn in partnership with others. If there is a commonality, the commonality is in the emphasis on fluidity, diversity, and the apparent breakup of permanence in the formation of knowledge and individuality. From the new pedagogical notions of children who "construct their own knowledge," to ideas about the reflective teacher and classrooms as "communities of discourse," identity is no longer understood in terms of universal norms of competence but in terms of norms that speak about the multiple and pragmatic actions through which individuals negotiate and construct knowledge. The contemporary meaning of partnership can be related to this notion of the actor/agent. At the same time, the dispositions and sensitivities that lie outside this normalcy form the "other," the capabilities and capacities of an individuality that is outside of reason itself and thus is disqualified for action and participation. The academic disciplines, new work order, and the military are homologous with an individuality, whose sensitivities and dispositions are produced in the practices of the reform programs of schooling. In each context is a range of cultural practices that relate to an individuality that works in contingent contexts of quickly changing information, who can process information and "problem-solve" flexibly in situations that change quickly.

Some Concluding Thoughts
The salvation themes in the current reforms are not merely paths for redemption but governing practices inscribed in the rules and standards of reason. I have sought to outline some of the contours of the cultural practices that demarcate the changing relationship between the pact and partnership. My concern in this analysis is with the rules and standards of reason in the different reforms as cultural practices that generate principles to order action and participation. My discussion initially focused on the systemic changes and the notion of community as interrelated and producing principles of governing that relate the pact and partnership—the collective and the individual. I then proceeded to consider the systematic practices as embedded within a field of cultural practices related to teacher education and pedagogical changes. The different practices were considered cultural as the amalgamation of those practices inscribe principles in which judgments are made, conclusions drawn, possibilities of change proposed, and the experiences and existence made administrable. The characteristics of a "problem-solving" child, the practices of educational psychology, the accountability measures of administration, and the characteristics of the teacher form an amalgamation of cultural practices. The significance of the practices is that they generate principles about the kinds of people we are and should be. Further, these principles also exclude as they include, and disqualify as they qualify individuals for action and participation.

The amalgamation or field is not one of policy intent or structural forces but of a historical phenomenon produced by its sets of relations.

While the salvation narrative of partnership is to empower the individual and remake the relation of the governed and the government, I have argued that the problem involves understanding such reforms relationally as problems of governing and the social administration of the individuality. Participation is not a normative principle that stands outside of its cultural practice but a governing principle formed and constituted through those cultural practices.

Programs of the partnership stand continually in relation to the pact to establish relations between the principles of individual action and the collective embodiment of the reason of the citizen who acts and participates in the name of liberty and freedom. Today's salvation stories are of an active sense of the lifelong learner whose emotional bonds and self-responsibility are circumscribed through networks of other individuals—the family, the locality, and the community. Life is a continuous course of personal responsibility and self-management of one's risks and destiny that occurs through "being" a problem-solver. But the lifelong learner is also a realignment of the governing principles of exclusion; to inscribe distinctions whose effects are to locate those who do not embody the characteristics of personal assessment of self-development and self-management as a lifelong learner.

Finally, I want to consider this discussion in relation to contemporary policy analyses. Much analyses locate the problem of reform in critiques of neoliberalism, or talk about a middle ground between conservative and social welfare policies, as the Third Way. This chapter suggests that policy analysis that uses such political frames of reference for critical analysis naturalizes the very historical questions of change that are in need of scrutiny. The social, cultural, and economic patterns that make neoliberalism or the Third Way plausible ways to reason about policy and change do not simply appear with the election of a new political party or leader, whether it is Thatcher and Reagan, or Clinton and Blair. While having different ideological configurations, the plausibility of neoliberalism and the Third Way are embedded in an amalgamation of cultural practices and social changes, to return to an earlier discussion, in which judgments are made, conclusions drawn, rectification proposed, and the fields of existence made manageable and predictable. Neoliberalism, for example, is not a "cause" but embodies a field of practices that relate collective images and narratives to the principles of action and participation. Reforms of partnerships, for example, circulate both in the Third Way and neoliberal policies as a double technology that order and classify who we are, should be, and who does not embody the distinctions and divisions of normalcy.

The chapter has excavated the rules and standards of reason that circulate in policy and research as an algorithm of governing the self. It has placed the salvation narratives of reform as linking the collective obligations of the state (the pact) with the characteristics of individuality (the partnerships). Partnerships had a duality: a focus on specific programs of decentralization and historically produced principles that govern the objects of reflection and action. This focus on policy and research is not to renounce salvation themes but to understand how they are instituted within the concrete practices of schooling. As Wagner (1994) writes, "the history of modernity cannot simply be written in terms of increasing autonomy and democracy, but rather in terms of changing notions of the substantive foundations of self-realization and of shifting emphases between individualized enablements and public/collective capabilities" (p. xiv).

Notes

1. This chapter was originally presented at "Educational Organizations in the Neoliberal Society," sponsored by the Interuniversity Congress of Organization of Educational Institutions, Granada, Spain, December 18, 2000. I wish to thank Miguel Pereyra, Ruth Gustafson, Amy Sosnouski, and Dar Weyenberg, for their comments on earlier drafts.
2. See discussion in introduction and also Nóvoa, A. (2002).
3. While there are different ideological concerns among current reforms, my argument will be that different ideological stances inscribe a frame of reference or system of reason at the level of concrete practices of teaching and learning. Thus, it is possible for political parties to change (from Clinton to Bush, or Thatcher to Blair), but the system of reason through which education is fashioned and shaped remains intact. This occurs, as I and others have argued, when the conditions of knowledge that produce its subjects go unexamined.
4. This study is entitled Educational Governance, Social Integration and Exclusion (EGSIE), sponsored by the European Union. It involves case studies in Greece, Sweden, Finland, Spain, England and Scotland, Iceland, Germany, and Portugal. See Lindblad and Popkewitz, 2001.
5. This idealized and romantic notion of community tends to leave out the systems of exclusion that went along with the traditions and rituals of participation.
6. It is not often realized in the United States that Clinton used this term to talk about his administration but with different rhetorics. The elder Bush as president and Clinton as head of the Governor's Conference set some of the agendas in the America 2000 Goals. The same general system of reason about school reform is maintained within the current Bush administration, which helps to understand the support of liberal Congressional Democrats of

Republican initiatives in 2001 in federal legislation concerning school testing and reading.
7. The issues of inclusion/exclusion are discussed in Popkewitz, 1998b; and Popkewitz and Lindblad, 2000.
8. If I can use an example of this in mathematics education, the constructivist literature continually cites the work of Walkerdine (1988) to register some limitations. Walkerdine studied how progressive, child-centered pedagogies were presented as universal rules of thinking and reasoning but were related to particular gendered and bourgeois mentalities that produced distinctions and divisions among children. The mathematical educational literature ignores this critique of universalizing reason in its presentation of constructivism as a universal.
9. I need to reiterate that while there are social and geographical distinctions between rural and urban contexts, the discourses of teaching carry those distinctions of place into the cultural practices of pedagogy. This is an "empirical" observation of the study.
10. The changes occurring are not merely those of the contexts described but involve simultaneously a globalization and localization that requires a more elaborate discussion than is possible here.
11. The different strands concern how knowledge is socially constructed, in some cases drawing on anthropological and sociological perspectives and others from psychology (one can compare the sociology of Pierre Bourdieu with the psychology of Howard Gardner). But as Bloor (1997) argues, the epistemological changes that relate to constructivism have multiple intellectual trajectories and no clear definition.

References

Alexiadou, N., Lawn, M., and Ozga, J. (2002). "Educational Governance and Social Integration/Exclusion: The Cases of Scotland and England within the UK." In S. Lindblad and T.S. Popkewitz (Eds.), *Education Governance and Social Integration and Exclusion: Studies in the Powers of Reason and the Reasons of Power. A Report from the EGSIE Project* (pp. 261–298). Uppsala Reports on Education 39. Uppsala, Sweden: Department of Education, Uppsala University.

American Council on Education. (1999). *To Touch the Future: Transforming the Way Teachers are Taught: An Action Agenda for College and University Presidents.* Washington, DC: American Council on Education.

Bhabha, Homi. (1994). *The Location of Culture.* New York: Routledge.

Bloor, D. (1997). "What is a Social Construct?" *Vest* 10/1: 9–22.

Bourdieu, P. (1979/1984). *Distinction: A Social Critique of the Judgment of Taste.* Cambridge: Harvard University Press.

Bourdieu, P. (1989/1996). *The State Nobility: Elite Schools in the Field of Power.* Stanford: Stanford University Press.

Butler, J. (1993). *Bodies that Matter: On the Discourse Limits of "Sex."* New York: Routledge.

Carnegie Council on Adolescent Development. (1995). *Great Transitions: Preparing Adolescents for a New Century*. New York: Author.
Chatterjee, P. (1993). *The Nation and its Fragments; Colonial and Postcolonial Histories*. Princeton: Princeton University Press.
Cherryholmes, C. (1999). *Reading Pragmatism*. New York: Teachers College Press.
Clifford, J. (1997). *Routes, Travel, and Translation in the late 20th Century*. Cambridge: Harvard University Press.
Cronon, W. (1996). "The Trouble with Wilderness; or, Getting Back to the Wrong Nature." In W. Cronon (Ed.), *Uncommon Ground: Rethinking the Human Place in Nature* (pp. 69–90). New York: W.W. Norton.
Danziger, K. (1990). *Constructing the Subject: Historical Origins of Psychological Research*. New York: Cambridge University Press.
Darling-Hammond, L. (1998). "Teachers and Teaching: Testing Policy Hypotheses from a National Commission Report." *The Educational Researcher*, 27, 4–10.
Donzelot, J. (1991). "Pleasure in Work." In G. Burchell, C. Gordon, and P. Miller (Eds.), *The Foucault Effect: Studies in Governmentality* (pp. 251–280). Chicago: University of Chicago Press.
Dussel, I., Birgin, A., and Tiramonti, G. (2000). "Decentralization and Recentralization in the Argentine Educational Reform: Reshaping Educational Policies in the 90s." In T. Popkewitz (Ed.), *Educational Knowledge: Changing Relationships Between the State, Civil Society and the Educational Community* (pp. 155–173). Albany: State of New York Press.
Fatis, S. (December 25, 1992). "Firms Trim Hierarchies, Empower Workers." *The Capital Times*, pp. 4b–5b.
Ferguson, R.A. (1997). *The American Enlightenment, 1750–1820*. Cambridge: Harvard University Press.
Foucault, M. (1979). "Governmentality." *Ideology and Consciousness*, 6, 5–22.
Gee, J., Hull, G., and Lankshear, C. (1996). *The New Work Order: Behind the Language of the New Capitalism*. Boulder: Westview Press.
Giddens, A. (1998). *The Third Way: The Renewal of Social Democracy*. Malden: Polity Press.
Hacking, I. (1999). *The Social Construction of What?* Cambridge: Harvard University Press.
Herman, E. (1995). *The Romance of American Psychology: Political Culture in the Age of Experts*. Berkeley: University of California Press.
Hultqvist, K. (1998). "A History of the Present on Children's Welfare in Sweden: From Frobel to Present-Day Decentralization Projects." In T.S. Popkewitz and M. Brennan (Eds.), *Foucault's Challenge: Discourse, Knowledge, and Power in Education* (pp. 91–117). New York: Teachers College Press.
Ingólfur Ásgeir Jóhannesson, Gurún Geirdóttir, Gunnar E. Finnbogason, and Sigurón Mýdral. (2002). "Changes in Patterns of Educational Governance and Social Integration and Exclusion in Iceland at the Beginning of a New Millennium." In S. Lindblad and T.S. Popkewitz (Eds.), *Education Governance and Social Integration and Exclusion: Studies in the Powers of Reason and the Reasons of Power. A Report from the EGSIE Project* (pp. 205–230).

Uppsala Reports on Education 39. Uppsala: Department of Education, Uppsala University.

The International Labour Organization. (1994). *Consequences of Structural Adjustment for Employment, Training, Further Training, and Retraining in the Metal Trades* (Report II). Geneva: International Labour Office, Sectoral Activities Programme.

Kohli, W. (Ed.). (1995). *Critical Conversations in Philosophy of Education*. New York: Routledge.

Latour, B. (1999). *Pandora's Hope Essays on the Reality of Science Studies*. Cambridge: Harvard University Press.

Lindbald, S. and Popkewitz, T.S. (Eds.) (2001). *Education Governance and Social Integration and Exclusion: Studies in the Powers of Reason and the Reasons of Power. A Report from the EGSIE Project*. Uppsala Report on Education 39. Uppsala, Sweden: Department of Education, Uppsala University.

Lindblad, S., Lundahl, L. and Zackari, G. (2001). "Sweden: Increased Inequalities-Increased Stress on Individual Agency." In S. Lindblad and T.S. Popkewitz (Eds.), *Education, Governance and Social Integration and Exclusion: Studies in the Powers of Reason and the Reasons of Power. A Report from the EGSIE Project* (pp. 299–329). Uppsala Report on Education 39. Uppsala: Department of Education, Uppsala University.

Malloy, J. (1999). "What Makes a State and Advocacy Structure Effective? Conflicts Between Bureaucratic and Social Movement Criteria." *Governance: An International Journal of Policy and Administration*, 12, 267–288.

Marshall, J. (October 2000). "From Colonialism to Globalisation: The March of Positivism in New Zealand Education." Paper presented at the Research Community: Philosophy and History of the Discipline of Education; Evaluation and Evolution of the Criteria for Educational Research Conference. University of Leuven, Belgium.

McEneaney, E. (2003). "Elements of a Contemporary Primary School Science." In G.S. Drori, J.W. Meyer, F.O. Ramirez, and E. Schofer (Eds.), *Science in the Modern World Polity: Institutionalization and Globalization* (pp. 136–154). Stanford: Stanford University Press.

McLaren, P. and Giardelli, J. (1995). *Critical Theory and Educational Research*. Albany: State University of New York Press.

Meyer, J., Boli, J., Thomas, G., and Ramirez, F. (1997). "World Society and the Nation-State." *American Journal of Sociology*, 103, 144–181.

Mirón, L. (1996). *The Social Construction of Urban Schooling: Situating the Crisis*. Cresshill: Hampton Press.

National Council of Teachers of Mathematics. (2000). *Principles and Standards for School Mathematics*. Reston: Author.

Noble, D. (1991). *The Classroom Arsenal*. London: The Falmer Press.

Nóvoa, A. (2002). "Ways of Thinking about Education in Europe." In N. Nóvoa and M. Lawn (Eds.), *Fabricating Europe: The Formation of an Educational Space* (pp. 131–156). Dordrecht: Kluwer Publishers.

Ohanian, S. (1999). *One Size Fits Few: The Folly of Educational Standards*. Portsmouth: Heinemann.

Popkewitz, T. (1991). *A Political Sociology of Educational Reform: Power/Knowledge in Teaching, Teacher Education, and Research*. New York: Teachers College Press.
Popkewitz, T. (1996). "Rethinking Decentralization and the State/Civil Society Distinctions: The State as a Problematic of Governing." *Journal of Educational Policy*, 11, 27–51.
Popkewitz, T. (1997). "A Changing Terrain of Knowledge and Power: A Social Epistemology of Educational Research." *Educational Researcher*, 26, 1–12.
Popkewitz, T. (1998a). "The Culture of Redemption and the Administration of Freedom in Educational Research." *Review of Educational Research*, 68, 1–34.
Popkewitz, T. (1998b). *Struggling for the Soul: The Politics of Education and the Construction of the Teacher*. New York: Teachers College Press.
Popkewitz, T. (in press). "Standards and Making the Citizen Legible." *Journal of the Learning Sciences*.
Popkewitz, T. and Brennan, M. (1998). *Foucault's Challenge: Discourse, Knowledge and Power in Education*. New York: Teachers College Press.
Popkewitz, T.S. and Lindblad, S. (2000). "Educational Governance and Social Inclusion and Exclusion: Some Conceptual Difficulties and Problematics in Policy and Research." *Discourse*, 21, 5–54.
Ravitch, D. (1996). *National Standards in American Education: A Citizen's Guide*. Brookings Institute.
Rose, N. (1989). *Governing the Soul*. New York: Routledge, Chapman & Hall.
Rose, N. (1999). *Powers of Freedom: Reframing Political Thought*. Cambridge: Cambridge University Press.
Rose, N. and Miller, P. (1992). "Political Power Beyond the State: Problematics of Government." *British Journal of Sociology*, 43, 173–205.
Russell, S. (1999). "Mathematical Reasoning in the Elementary Grades." In L. Stiff and F. Curcio (Eds.), *Developing Mathematical Reasoning in Grades k-12* (pp. 1–12). Reston: National Council of Teachers of Mathematics.
Schram, S.F. and Neisser, P.T. (Eds.). (1997). *Tales of the State: Narrative in Contemporary U.S. Politics and Public Policy*. Totowa: Rowman & Littlefield.
Simola, H., Rinne, R., and Kivirauma, J. (2001). "Shifting Responsibilities, Insolvent Clients and Double-Bound Teachers—the Appearance of a New System of Reason in Constructing Educational Governance and Social Exclusion/Inclusion in Finland?" In S. Lindblad and T.S. Popkewitz (Eds.), *Education Governance and Social Integration and Exclusion: Studies in the Powers of Reason and the Reasons of Power. A Report from the EGSIE Project* (pp. 59–97). Uppsala Reports on Education 39. Uppsala, Sweden: Department of Education, Uppsala University.
Smith, M. and O'Day, J. (1990). "Systemic School Reform." *Politics of Education Association Yearbook*, 233–267.
Smith, R. (October, 2000). "Education, Truth and Hunger." Paper presented at the Research Community: Philosophy and History of the Discipline of Education; Evaluation and Evolution of the Criteria for Educational Research Conference. University of Leuven, Belgium.
Stone, L. (1999). "Reconstructing Dewey's Critical Philosophy: Toward a Literary Pragmatist Criticism." In T. Popkewitz and L. Fendler (Eds.),

Critical Theories in Education: Changing Terrains of Knowledge and Politics (pp. 209–229). New York: Routledge.

Van Bendegem, J.P. (1996). "The Popularization of Mathematics or the Pop-Music of the Spheres." *Communication & Cognition*, 29, 215–238.

Wagner, P. (1994). *The Sociology of Modernity*. New York: Routledge.

Walkerdine, V. (1988). *The Mastery of Reason: Cognitive Development and the Production of Rationality*. London: Routledge.

Zeichner, K. (1996). "Teacher as Reflective Practitioners and the Democratization of School Reform." In K. Zeichner, S. Melnick, and M. Gomez (Eds.), *Currents of Reform in Preservice Teacher Education* (pp. 199–214). New York: Teachers College Press.

CHAPTER 2

PARTNERSHIP AS A FLOATING AND
EMPTY SIGNIFIER WITHIN EDUCATIONAL
POLICIES: THE MEXICAN CASE

R. Buenfil Burgos

Introduction

Partnership is a word that can be associated with the idea of corporation, company, firm, business, legal relation involving rights and duties, joint venture, participation, close cooperation, and collaboration. It frequently appears in national and particularly in educational policies resembling a universal—almost a natural—value upon which all specific agents and governments in general, would agree. However, the meaning of this word changes in each particular moment and site of enunciation. This entails the examination of two discursive political operations: on the one hand, circulation and partial fixation; on the other, ambiguity and political moves.

The *signifier*[1] partnership flows and circulates throughout a variety of meanings, and sites of enunciation and it has become what it is today through a series of discursive articulations throughout history. The argument I will sustain in this chapter is that partnership grants some temporary fixation to the flow of meaning in educational policies; thus it operates as a *nodal point*[2] and it can be understood as a *floating and empty signifier*.[3]

The ambiguity of the signifier is generally understood as something undesired, that is, the vague and nonspecific character that should be amended such that it could become clear and precise. On the contrary, I will argue that the ambiguous character of a signifier is politically productive, since if it were possible to definitively stop the flow of meanings and then claim that one and only one is the right meaning, then there would be no possibility to political action, dissention, negotiation, and hegemonic practices. I will argue that in Mexican educational reforms, the constitutive ambiguity of partnership enables it to mean both the inclusion of civilian (i.e., nonofficial and private[4]) involvement in educational decisions and the withdrawal of the government's responsibility in public educational matters.

I will analyze how partnership is constructed within recent educational policies in Mexico, namely how it is linked with political progress and democracy[5] throughout its various enunciations in official documents. To these ends I will benefit from a deconstructive gesture and a genealogical move.[6]

The way in which partnership is construed has political and ethical consequences that may appear overlooked in the rhetorical construction of reforms. I will briefly discuss these consequences from a discourse theory and political analysis perspective (Laclau and Mouffe, 1985; Laclau, 1990, also known as *discourse political analysis* Buenfil, 2000a).

I will start by presenting the meanings fused in the signifier partnership within contemporary schooling reforms and laws, and the way in which they circulate throughout other branch programs and across national and international discursive series, sites of enunciation and of course, different and even opposed cultural systems. This initial presentation will already involve basic analytical remarks (e.g., area of dispersion, ambiguity) to gradually set the grounds for my further argument. In the second section I will try two analytical movements: deconstruction and genealogy to account for the trajectory and operations performed by the signifier. Finally, I will discuss some political and ethical angles of the ambiguity inherent to the signifier partnership in this new horizon of plenitude and salvation[7] promised by our post-welfare global condition.

Partnership as a Component of a Broader Discursive System: The International Agencies

Educational discourses issued by international agencies constitute a specific *language game* (Wittgenstein, 1963)[8] within which partnership is constructed and this enables us to understand through relational operations a different dimension of this signifier as a key component of Mexican educational reforms. In the following lines I will draw upon the meanings conferred on partnership in three different discursive configurations: the World Bank (WB), the International Monetary Fund (IMF), and UNESCO (March 1990).

A brief review of WB's value configuration shows us also an unbalanced construction of partnership where its benefits prevail over its risks. The WB in its report on education declares that partnership is a key value insofar as it is linked with the involvement and commitment civilians will have vis-à-vis school management (Banco Mundial, 1996). After diagnosing the state of schooling on the planet, the WB states a series of recommendations. Amongst the six priorities they set for educational reforms, partnership has the fifth place. As it is an axis of the whole proposal, an entire chapter is devoted to display the features and contours of partnerships. In chapter 9, grounding their interpellation on world experiences, parents and the whole community are called to collaborate in school management and in financing teaching, through the idea that consumers can demand of suppliers some goods and their quality, and therefore can be involved in decision-making. This has been sometimes limited because governments do not always allow private schooling. The WB considers some risks when calling for partnerships, namely the lack of information parents may have.

Three articulated logics are exhibited in these statements: administration, consumption, and the contours of public responsibility upon schooling. This is the case since the idea that parents must intervene in financing, demanding the goods, and making decisions about school management, refers to the fragile frontier between public and private accountability.

UNESCO's (1990) discourse is structured around basic competencies, partnership, and life quality. Partnership is constructed as the responsibility over education assumed by both nongovernment agencies and government establishments, private groups and educational agents—schoolteachers, parents, students, school authorities.

In 1999 the IMF together with the WB submitted a new proposal presenting partnerships as one of the three capital features of their strategies against poverty. It is associated with the involvement of civil society (the poor included), select agencies, foundations, and financial agencies. It is constructed as a means to an end (e.g., plans, follow-up implementation and improvements of the strategies against poverty) and also as the coordination of WB and IMF collaboration. It is a feature of a process involving a common understanding, analysis, results, and outlook of the way to fight poverty, and also a compulsory attribute of the strategy for which WB and IMF credits are asked. Partnership is the *sine qua non* for any thinkable and presentable program against poverty and operates as the *nodal point* according to which the program will be evaluated. It is also indicated that multilateral agencies and regional banks should partake. Finally, the highest hopes of the benefits achievable through these strategies against poverty rely precisely in the participative quality demanded in the process (Fondo Monetario Internacional-Asociaiación Internacional de Fomento, 1999).

Considering the relational character of any identity, examples will be provided whereby one can observe the displacements of the signifier partnership throughout different discrete discursive configurations (e.g., domestic educational plan, the solidarity program, and so on) and national plans; and also their circulation (or traveling) along national and international proposals (such as the UNESCO's, WB's, and IMF's). The condensation of several meanings in each enunciation will also appear attached to the same signifier. Thus partnership as part of the configuration of meanings within educational reforms can be easily traced, analyzed, and interpreted in the light of its iterability[9] along these other discursive series.

Partnership in Mexican Schooling Policies Today

Some remarks may serve as the background to contextualize the meanings of partnership in the twentieth century, tracing back in history some key moments and educational reforms.

- Partnership was constructed as a key means for the political organization of the masses within the *socialist education* (1934–1940), (Buenfil, 1990 and 2000b).
- Partnership was later constructed as the participation of the private and ecclesiastic groups in educational matters and in general terms, the involvement of businessmen in public economic affairs in *education for national unity, for love, and for international peace and solidarity* (1940–1946, World War II days) within a welfare state imaginary.[10]
- Constructions of partnership within *educational modernization* reform (1988–1994) are inscribed into a nonreversal process of decline of the welfare state imaginary (this will be explored in the following).

The following exercise displays the signifier's impressive area of dispersion that is visible in two interwoven discursive series belonging to different societal scales: firstly, national projects such as plans for national development and a branch program; and secondly, some recent schooling programs. I will then present the laws concerning education. This exercise will help to show the massive proliferation of signifieds linked with the signifier partnership within (national and international) official sites of enunciation. It will also show how the same themes (or equivalent meanings) reappear, and similar links with other signifiers are constructed in the attempt of filling in the emptiness and fixating the field of educational partnership.[11]

Partnership as a Component of a Domestic Discursive Ordering: National Plans for Development

Educational reforms, programs, and laws acquire meaning insofar as they are part of broader systems. In this discursive scale numerous iterations of the signifier partnership were found along with such synonyms as co-responsibility, initiative, and volunteering namely in the educational heading of the national development plan of 2001–2006 and also in the previous plans of 1995–2000 and 1989–1994. Calls to civilian partnerships (parents and other private groups) were also constructed as a goal linked with democracy, educational quality, equity, and modernization (Poder Ejecutivo Federal, 1989). The paragraphs devoted to education also mentioned partnership as the input schoolteachers supplied for what had been achieved and as a necessary means to attain *educational modernization*. In more than 280 pages, this report repeatedly enounces partnership (180 iterations, see Secretaría de Desarrollo Social, 1995).

The signifier partnership plays a structuring role, as a crucial piece to legitimize the popular, social, and consensual character of these programs. It is constructed as:

- the equitable distribution of responsibility upon education between public and private sectors;
- a consequence of the democratic regime that would in turn, reinforce partnership, lead to freedom and guarantee full democratic development;
- a citizen demand inspired in national values that has now been achieved: social and civilian associations that take place within normative and legal frames;
- a link with the frontiers between public and private, with citizen involvement and political culture, with citizen commitment in public programs and public administration;
- an improvement of public services—education included—carried out by parents, private schooling, businessman, government, authorities, and the whole community. It is connected with financing, planning, and complying diverse programs.

Partnership in the Frame of Educational Reforms
The national education program of 2001–2006 can be understood only within the wider frame recommendations of international agencies and the national development programs (supra). This relational character of this educational policy (and indeed of any other identity) enables us to explore the forms and sites through which this values displace themselves (or in Popkewitz terms, *travel* around) from one enunciation to another, and the ways in which these represent much more than the single word in which they *fuse* different meanings. In other terms, this relational character allows us to interpret the *displacements* and *condensations* of key values such as partnership, globalization, equity, and so on.[12] In the national education program 2001–2006 partnership (in Spanish *participación*)[13] appears not less than 140 times (in a 266-pages document, see Secretaría de Educación Pública, 2001).

This proliferation of meanings can be traced in previous educational programs: from 1988 to 1994 and from 1994 to 2000. *Modernization* was the key notion organizing social, economic, and political reform. In the educational modernization program the signifier partnership appears 120 times (Consejo Nacional Técnico de la Educación, 1992). Numbers alone are never enough to understand the meanings and roles played by this signifier. So now let me unpack it and by intertwining the previous discursive series, show the impressive area of dispersion of this signifier crisscrossing the four national and four educational plans mentioned earlier.

In administrative terms it is linked to collective educational management, planning and evaluation, federalism and decentralization; economic data and financial strategies; consultation to, coordination and partaking with all social sectors; political forms; as a transparent, lawful, honest and, efficient management. Considering political and civic values, it is constructed as equivalent to democracy, pluralism, civilian values, national identity and a cultural approach, gender equity, society development and progress, collective commitment; solidarity, teamwork contribution; a fresh and creative form of social cooperation, collaboration amongst authorities, teachers, students, parents, and community in the solution of school problems; political and labor opportunities and involvement, students partaking in school issues. Involvement, rights and duties, and practices comprise the following:

- Social involvement: communitarian commitment, collective will, social welfare, calling the whole society to be involved in all sorts of activities: from public decision-making to neighborhood surveillance; it ranges from equality concerning gender, ethnic,

generational (childhood, adolescence, adulthood, and third age) and cultural inclusion to active involvement in some political activities; from volunteering in social activities to engaging in small enterprise and self-employment.
- Rights and duties: parents rights; youth electoral and political rights and capacity, solidarity, participation, collective enterprise, collective benefit, progress, indigenous culture, and university students' moral duty.
- Actions such as: work, participation with economic resources, the supply of regional goods, surveillance over its management, direct communitarian action, schools maintenance and refurbishing, and design and execution.

In view of educational matters, it is associated with a cognitive value, a learning feature, curricular strategies, evaluation (one of the four brand new elements of these programs):

- a partaking disposition amongst teachers, amongst educational institutions; educational and cultural, media and research establishments, and amongst schools and society concerning inspection and accountability.
- schooling quality concerning formal and nonformal (incidental) education; basic schooling (kindergarten, primary, and secondary cycles), high school; lifelong learning, adult education (closely related with other social programs), with teacher training and updating and labor training. It also concerns preuniversity (upper high school) and technological options, where not only parents but especially businessmen are called to get involved. Higher education also is organized around few lines: partnership, evaluation, and fees that draw the contours of technological qualification, open university, scientific, humanistic, and technological research; and finally, with the equipment and maintenance of infrastructure.

In argumentative terms[14] partnership is constructed in the following ways: as a background of today's changes, as a tradition in national idiosyncrasy, a form of life, a condition for democracy, a key to meet social needs, a political strategy, an action to be promoted; as a solidarity distinctive feature, as the education axis, ground, and principle; in terms of community–government collaboration, as a social and plural mobilization:

- as an end in itself to be pursued (democracy, equity, gender claims, indigenous demands, and so on), and a means to achieve a further

end (e.g., a means to get political transition, to a new society, to educational reform, to teachers updating); the strengthening of public higher education, everybody's commitment to education; as a means to achieve educational quality, the compliance of normativity, and decentralization;
- as an opportunity to take advantage of, as a requirement of our global condition, and as a challenge to face;
- as the cause of a consequence (e.g., the foundation of a Council for National Partnership on Education); as a consequence of an act, as consequences of our history, and as the practice whose consequence is the plan goal; as an act with positive consequences and as a consequence of a form of government;
- as a value to be achieved and as a strategy to achieve other values.

Partnership is linked to different topics: education, schooling infrastructure, economic and social intervention, and impact, social assistance collaboration and participation, inclusion of the retired teachers, health services, urbanization and development services (for both cities and rural areas), acknowledgment, counseling and legal advise to young people and the community, food supplies and rural stores, indigenous goods production (coffee, arts and crafts), small business and small property; culture and justice (anthropology and law); popular banking; women's access to services, environmental sustainability, municipality and regional management, roads and highways, and federation and national zones development programs inter alia. It is intertwined with economic transition and global technologic changes. It is associated with social changes such as women labor involvement, concern for the marginalized (living conditions, health, welfare, security), with the duty the whole nation has concerning education, with a human feature to be achieved through education but also with an attribute that education itself should reach something the government expects from the population; a structural part of educational challenges and changes in the short and the long term, or a subject of a law. Partnership is constructed as opposed to:

- government-only management, government-only financed service, centralization, obstacles to parents collaboration, limits to private intervention, extreme poverty, discrimination, inequality, corruption, lack of democracy, and all sorts of limits and obstacles to gender equity, ethnic inclusion, and so on.
- recession, stagnation, decline, abandon, corruption, idleness, indolence, selfishness, inefficiency, turbid management, government monopoly, and centralized administration.

- society stagnation, lack of involvement, poverty, delay, lack of responsibility, authoritarianism, lack of democracy; inefficiency in or lack of consultation to and partaking with all social sectors; unilateral management, centralism, corruption and lack of accountability, and feeble participation of society in educational matters.

The public and the private appear interwoven with public schooling partnerships, which are construed as opportunities for personal development. Parents and private groups are called to be partners of the educational enterprise in order to promote financing, management, and teaching activities as well as in incidental education. Of special interest is how the private–public relation is presented under the light of partnerships as the erasure of all political distinctions between public and private schooling. This is the case because both have to be partners in the building of a future Mexico. There is an insistent appeal on parents and private sectors.

Partnership in the Frame of Educational Laws

Another context of enunciation is constituted by the laws dealing with education. Many nodal points can be located to understand the trajectory of participation in this terrain: general law of education (1993), *organic laws* (1941 and 1935), and Article III of the Mexican constitution, which deals with education and has been transformed approximately within the same periods. Continuities and discontinuities amongst these pieces of legislation, namely Article III but also the organic and general laws, considering the focus of this chapter, show this:

- In the laws of the 1930s and the 1940s laws, the overall idea of a strong welfare state that would guarantee public services for the population still was a strong imaginary of plenitude. In the 1990s law, traces of the welfare state decline are evident and its compensation with calls on private and civilian partnerships becomes frequent, promising salvation through their intervention in public schooling.
- Participation in the 1930s law is rather linked with a political strategy to get organized and with the distribution of rights and obligations amongst Federation, states, and municipalities; parents, tutors, teachers, and authorities. In the 1940s law, partnership is connected with the corporate support of official policies, and, like

in the previous law, with the distribution of rights and obligations amongst social agencies. In the 1990s law, partnership is linked with the idea of liberalism, pluralism, and autonomy from the government monopoly on public schooling; thus private and ecclesiastic involvement in decisions over public educational matters becomes formally legitimized and the calls on parents and the community as a whole to collaborate with the school come to be more explicit.

The general law of education was reformed in 1993 in accordance with the policy issued in 1989 and includes some subtle and some dramatic changes concerning the private and religious capacity of decision-making in public educational matters. To start with, it is interesting to note that for the first time in the history of this law a whole article is devoted to designing and establishing the grounds for nongovernmental partnership in public schooling. Article 7 deals with the idea of social partnerships in education involving parents, social partnership council, and the media. In previous laws (1935 and 1942) participation was basically linked with the coordination of educational services amongst federation, states, and municipalities or teachers, and authorities and parents. Partnership is construed in this law as a result of a democratic form of government, as the outcome of a correct educational planning; and the establishment of strategies for the constitution and financing of partnership councils. The signifier is connected with social roles and interests and with *solidarity*, provided that through partnership private groups and the community as a whole would collaborate in financial and administrative terms. This signifier is also linked with volunteering and the altruistic cooperation of private and civilian agents and the population at large.

Considering the reform documents and the laws previously mentioned, the signifier partnership not only is excessively reiterated within the official documents concerning the schooling system, but it also exhibits a high degree of dissemination since in each *iteration* it puts on view both the reappearance of the same themes and a variety of possible meanings. The forms in which it is constructed indicates the structuring role it plays vis-à-vis other social programs and other signifiers that remain to be treated.

After this extended exercise and reading of this discursive explosion and exponential increase—to use a mathematical metaphor—of the signifier partnership in official documents, namely the reiteration of its benefits in short and long terms, little doubt can remain about its wide area of dispersion and its being a nodal point in the articulation of a new promise of happiness and plenitude, the new salvation horizon constituted by globalism, partnership, marketing, and the rule of civilization.

How did Partnership Come to be What it "is" Today?

Two analytical moves will be attempted in this section: a deconstructive reading (Derrida, 1997) to examine some political and semiological movements of the signifier as it is exhibited in today's educational programs, and a genealogical exploration (Foucault, 1977) to investigate how partnership became what it is now.

In a deconstructive gesture one can read the way in which a binary opposition is constructed when both the international agencies and the Mexican programs today produce an implicit rejection of government-only managed schooling, and suggest that they are a non-democratic government monopoly. One can seek the value hierarchy structuring these discursive constructions of partnership: there is one valuable way to administer schools and this comprises parent and other private sector involvement. Beyond this basic agreement, abundant dissemination has been presented earlier in terms of the variety of topics and argumentative constructions (e.g., exercise supra); the differences between financial and administrative partnership, parents and community partnership; and when, for instance, the WB explicates the good consequences and the risks of ill-informed partnerships.

No matter how distant the WB and UNESCO may be in terms of goals, the countries each one represents, intervention strategies, and actual influence over nations, some similarities concerning their constructions of partnership can be easily found. No matter how different in influence the national development program and the educational plan may be, the same logic is present in both and the structuring role played by partnership is visible in both.

The *universal* and *quasi-natural* character of the signifier partnership in national programs is an effect of the erasure of the institution of an exclusionary regime that ruled out previous meanings or removed their structuring status. Three aspects are involved within the post-foundationalist[15] perspective I sustain. First, there is no room for a transcendental and a-prioristic universality, but rather for a universality that is an outcome of political relations and hegemonic practices (Laclau and Mouffe, 1985, Laclau, 1996).[16] Thus a universal idea (plan or agenda) is a particular one that at a certain point has articulated others around itself, has constructed political frontiers thus antagonizing other ideas (i.e., it is a particular value that has hegemonized a discursive field). Therefore, universality is always contaminated by particularity (and vice versa). Second, a deconstructive reading enables us to see any meaning fixation as an effect of the erasure of a moment of decision by means of which something has been included and something excluded (Derrida, 1982). The inscription in a text or if one so wishes, the moment of inclusion and exclusion that defines any discursive system tends to be blurred by repetition and sedimentation thus concealing its political character and rendering "natural" the resulting structure (Derrida, 1982). *Mutatis mutandis*, in Foucault's terms, a "normal" regime (Dreyfus and Rabinow, 1983) is produced through *dispositifs* (involving both linguistic and non linguistic practices) whereby images of what is right, proper, beautiful, smart, just, democratic, and what is not are inscribed in everyday life (i.e., the emergence of an exclusionary regime, Foucault, 1977). Third, universalization, naturalization, and normalization involve a political moment by means of which a system is defined, where boundaries separate what is included and what is excluded; this moment is erased and the resulting regime appears as if it were given as such, has always been there, and so on.

One possibility to seek the course of the values articulated around partnership consists of tracing back in history where they came from, and highlighting what was included and what was left behind in this trajectory (*supra* on educational laws). Three key ancestors of partnership may be recognized in the first half of the twentieth century. First, there was a precursor from the general governmental postrevolutionary agenda. In the 1930s, this signifier was attached to official programs as a call to political grouping and labor organization and produced two exclusionary systems where partnership was construed in antagonistic terms (mutual negation).

- The government construed partnerships as political organizations of the popular sectors leading to a socialist democracy, as a form for the distribution of rights and duties amongst federation, state governments, municipalities, and so on, and as the participation of parents and private groups under strict federal government rule.
- The opposition construed partnerships as the inclusion of the private groups to replace the federal government monopoly, also as a form for the distribution of rights and duties amongst federal state, local governments, municipalities, and so on.

Second was the one emerging within the national development program and defining educational plans. It can be found in technological training schemes for both upper high school and higher education (i.e., National Polytechnic Institute).[17] The formerly launched welfare state was still legitimate as a promise of plenitude (i.e., an imaginary horizon); but undermined by a liberal economy and politics, and a conservative and prejudiced morality and educational program. This discursive reconstitution produced a different exclusionary system in the 1940s.

- Both the construction of the government and the nongovernmental agencies became one: partnership came to be the inclusion of the private groups to replace the federal government "monopoly," as a form for the distribution of rights and duties amongst federal state, local governments, municipalities, and so on.
- A political antagonism was constructed with the previous regime and their "alien" socialist notion of partnership as political organization.

And third was the one produced by the ecclesiastic opposition to the government and its general and educational programs.

- Partnership then consisted of the participation, involvement, and inclusion of the private groups mainly organized by the Catholic church (parents associations, private confessional schools, religious congregations, and so on), and businessmen.
- Partnership was antagonistically construed as the formal exclusion and rhetorical discredit of the alleged government monopoly over public education.

This idea of partnership as the involvement of business, and especially of factories, in technological training was preserved—though without success—for many administrations. The partnership of industries with actual equipment and "real" machinery was supposed to aid technological development. In the 1980s this signification through iteration became sedimented and reached a new climax as a successful program (Bernal, 1998).

By this genealogical gesture, it is possible to see the displacements of the signifier along multiple sites of enunciation (e.g., programs and laws), its condensations in different discourses (reforms and recommendations), and by this means its numerous resignifications (road to salvation, administrative strategies, and so on). Successive inclusions generate a wide area of meaning dispersion and gradual exclusions reduce this area to a particular set of signifieds (either as a panacea or as the universal signifier of all evils) thus universalizing, normalizing, and naturalizing it as a signifier.

However, beyond specific technological training, I want to stress that our contemporary (beginning of the twenty-first century) exclusionary system is globally interconnected, postindustrial, post-welfare, post–Cold War, with different geopolitical frontiers, different social movements and agents, and an increasing suspicion of the Enlightenment imaginary[18] and a more precise and frequent visibility of the planetary unequal distribution of wealth. It exhibits new parameters for the discursive construction of partnership. In Mexico, this exclusionary system indeed resignifies partnerships under domestic conditions, however, not in a really different manner.

This process of inclusions and exclusions, namely the organized forms of persuasion through which partnership has become *natural*, involves the sedimentation of the "novel" value and the gradual obliteration of the moment of its institution occupying a prominent position within the legitimized configuration of educational and social values (i.e., the decision whereby these were included and others—welfare values—were excluded).

A genealogic move may help in my analysis to avoid the temptation of seeking the mythical origin of partnership as the philological or etymological essence identical to itself. Instead, I will track the previous exclusionary system where the signifier had a position and the turning point where this position was changed. I will therefore investigate some discursive series that have left a mark in today's policies: national idiosyncrasy, a tradition in national programs, domestic struggles in history (ecclesiastic demands for intervention, parents associations' fight over education intervention), intellectual and economic changes of a global imaginary, and international recommendations.

Partnership as a value, as a political legitimizing strategy, and as a cognitive learning strategy or as a target to achieve throughout the history of schooling reforms in Mexico and within national development plans, shows marks of the decline of welfare state values. It comes to be "natural" and "normal," "necessarily universal" by means of the condensation of all these elements, and their sedimentation and naturalization by means of the symbolic investment of these values on rituals, institutions, laws, and so on.

Traces of national idiosyncrasy can be found, especially in national programs, since already in pre-Columbian days, communitarian management was supported by collaboration. During the Spanish regimes also, participation of the nongovernmental agencies was frequent. We could find this idea of community participation in public affairs in different epochs of the nation's life.

Marks of community participation as a tradition in national programs can also be detected. This concept was already present in native pre-Columbian traditions, and was also enhanced during the Mexican Revolution, namely as a political form of organization and intervention during Cardenas's administration in the 1930s. These two traditions—the pre-Hispanic and the revolutionary—then produce a surplus value in national programs. Since then partnership does not appear merely as an international tendency but as something that was already present in Mexican forms of management. This certainly reinforces its interpellatory power.

Imprints of domestic struggles in history can be tracked in the discursive configuration of partnership. In the Mexican case, the Catholic church and parent societies systematically fought to recover the alleged government monopoly over education,[19] namely from the political reform (1857 onward), when the Catholic ecclesiastic power was removed from the government. The struggles these forces waged against each other also shows that in the alleged origins of a discursive formation rather than in the essence, the genealogist finds struggles (Foucault, 1977). Ecclesiastic political moves from the nineteenth century onward and parent federation demands during the twentieth century to open the space for intervention in public schooling, have also left a mark in the policies we witness today.

Traces of intellectual and economic changes of a global imaginary also permeate the partnership discourse. The welfare state imaginary, the geopolitical frontiers, and the ideals of centralized management are declining thus giving place to a liberal ideal, a new setting of geopolitical boundaries, ideals of decentralization, and so on, with the marketing logics permeating the whole scenario. The "thinning of the state apparatus" is increasingly legitimized which in turn, justifies the involvement—and sometimes obligation—of nongovernmental agents in public matters. In the twentieth century all these tendencies were less interconnected and their expanding effects were less visible. Today our global conditions enable multidirected contacts among nations, regions, hemispheres, and so on, accelerating, intensifying, and rendering these effects more perceptible on each other.

Tracks of the IMF, WB, and UNESCO urging educational partnerships are also perceptible in Mexican reform. A simplistic view would only see the domination and imposition exercised by these agencies over national programs. I partially acknowledge this, but I do not take it as the full account. There is also an interpellatory capacity that makes this urging a reasonable recommendation and this is not a minor issue. On the one hand, UNESCO does not stand for the same interests as the WB or the IMF—so you have the argument that even the monetary-guided and the educationally led agencies share the same view, *ergo*, "it must be right." On the other hand, this persuasive power increases with the features and history of the addressee (in our case, the Mexican government and local office and branch authorities).

All these traces, imprints, tracks, and inscriptions are endless attempts to fill in the emptiness of the signifier. As part of the conditions under which these discursive configurations operate, there have been numerous contestations to these post-welfare global advances. Many of them have been produced by well-intended left-wing intellectuals, some by liberation theologicians, and a few more by the new "globalophobic" militants.[20] These contestations usually involve the absolute rejection of globalization, neoliberalism, and the withdrawal of government responsibility over public services (e.g., health, schooling, and so on). This absolute rejection has a conceptual basis whereby globalization is understood as the *universalization* of the market-economy and the flux of investment capital, the ruling of all aspects of the inhabitant's life by imperialism and transnational capital, thus constituting a *homogenized* world capitalist society, and the exclusion of cultural, political, economic, and national identities of Asia, Africa, and Latin America (for a discussion, see Buenfil, 2000a). This genre of understanding operates as the broader frame within which partnership is constructed. Hereby, partnership is construed as the withdrawal of public services from the agenda of the neoliberal political system, as the government elusion of its responsibility to provide social goods for the population, what I have been calling the post-welfare state. The welfare state, then, is constructed as the origin of equity and the guarantee of salvation; therefore its withdrawal is construed as opposed to progress and social justice. This construction of partnership as a component of the contemporary "Pandora box" assembling all the evils in view (and underneath), is partly right since it is difficult to oversee the exploitation involved in imperialism and the unequal development it produces (in the former colonies). It would be clumsy and insensitive to ignore the living conditions of great proportions of the population of Asia, Africa, and Latin America and what the withdrawal of public services means for them.

However, I do not subscribe to this "Pandora box" position since in conceptual terms, it would be incompatible with what I have been arguing for thus far (the ambiguity of partnership, the open character of any discursive configuration, the historicity of all values, the political status of any discursive construction, and so on). I do not accept this construction because in analytical terms, it overlooks important dimensions of partnership—(for example, organization, visibility, self-management, some decision-making) and does not analyze the way in which the existing welfare state never managed to deliver the goods. Finally, in phenomenic or ontic terms, I also object to this demonized version of partnership because it does not acknowledge other meanings construed by the very popular sectors—they allege to speak for (e.g., collective commitment, mutual help, self-acknowledgment, self-awareness, pride and dignity, reduction of dependency and the habit to expect welfare assistance, and no more government charity).

Having in mind the two analytical moves previously exercised, one can understand on the one hand, the *naturalization* (Derrida) of the signifier partnership (i.e., the erasure of its political and discursive instauration, that is, its historicity), its dissemination and constitutive ambiguity; and on the other hand, one can also realize its *normalization* (Foucault) through the establishment of a different exclusionary system (post-welfare) and its sedimentation by means of legitimized dispositifs (laws, programs, reforms, budgets, and so on). Both *naturalization* and *normalization* are in this chapter understood as political constructions.[21]

Discussion

I will now try to interlace the main argumental threads I have displayed in previous pages in order to open my own position on partnership to further discussion. There is no essential or ultimate sense of partnership, neither its meaning as the withdrawal of the government responsibility over public education, nor as the means to liberal democracy and salvation; nor its signification within the market-economy discourse nor that related with a more democratic political organization. This signifier is, just as any other, an overdetermined (note 12) symbolic and historic construction. Partnership in Mexican educational programs is, in addition, the outcome of both the condensation of domestic (national and local) and international processes and the displacement of the signifier throughout different moments and sites of enunciation and discursive series.

Partnership is a nodal point, since it operates in a double movement: on the one hand, because it is a floating signifier in educational and national development plans. This means that it is available in several governing narratives and thus can be retrieved and articulated to particular meanings. And on the other hand, as we have witnessed, because it is a signifier that in spite of being highly sense-loaded, paradoxically its *universality* enables it to organize and grant some fixation to the flow of meaning available in these discursive configurations. On the contrary, it operates as the recipient of a variety of meanings (i.e., as a surface of inscription) thus preparing the terrain to become a *universal* signifier of democracy, progress, and salvation. Focusing a different angle one can also claim that this capacity of accepting numerous meanings—its *emptiness*—is the very condition for the *universality* of the partnership concept. Partnership has become a universal value in a discursive field widely organized around a promise of plenitude and salvation by values such as democracy, progress, globalism, marketing, productivity, and so on. The fact that this signifier and also other nodal points (such as equity) are filled with "content" and thus are constituted as practice in relation to particular forms of governing and a reconstitution of the state (Popkewitz, 2000) are political and discursive operations involved in the hegemonic practices under examination in this chapter. From the analysis and genealogy of partnership it is possible to say that this association far from being the result of necessary links is rather the result of contingent articulations that normalized this representation as the road to salvation, erasing the very moments of inclusion and exclusion and its undecidable character, and presenting it as the only possible representation of plenitude.

Once we move in the terrain of undecidable structures (Derrida, 1982) we can realize that this open character is the condition for decision.[22] At this point the room is open to political and ethical[23] moves within this undecidable structure—these moves have been explored elsewhere (Buenfil, 2000a; Ruiz, 2000)—which is precisely why decisions over salvation plans, partnership, and so on can make the difference in the implementation of these plans.

The generalized hegemony—antagonisms and articulations, inclusion and exclusion—of our globalized post-welfare state condition and its peculiar promise of plenitude and salvation, can hardly be denied. However, if one conceives globalization not as mere imperialization and domination of an economic, political, and military viewpoint, but as the interconnection of unequal, productive, and conflictive values and forces,[24] then there is no reason for a pessimistic approach, but rather for a politically involved stance. On the one hand, salvation narratives have always been available though structured around different nodal points: the divine, the secular, in its various versions (Enlightenment amongst the most powerful, the Marxist, the neoliberal, and so on).[25]

On the other hand, contemporary salvation narratives are also ruled by an aporetic logic—they are both necessary and impossible—and, just as any other narrative, they are incomplete and fissured; *ergo*, they are open to resignification, contestation, dislocation, and subversion. Considering these circumstances, the post-welfare condition that is embedded in the filling of meanings that occurs through the mobilization of multiple discourses into the space of partnership, (even if a pessimistic person equates it with the Pandora box) there is always a possibility to resignify this value.

The ambiguous character of the signifier partnership makes visible that it means both the road to salvation—to democracy and national progress, equity and control over corruption, the civil society involvement in education, and so on—and also the road to imperialist exploitation—that is, the withdrawal of government responsibility over public education, the handing over of education to private management. This exhibits the way in which this ambiguity allows decisions to be made and demands positions to be held.[26] This also makes visible how partnership *qua* nodal point is articulated with salvation narratives thus making possible some conditions of governing and some success in the state enterprise to define "the" representation of a nation's self-realization, progress, democracy, in short, plenitude.

In addition, partnership as part of the global exchange, is for some today the pandora's box and is not immune to heterogeneous micro-contestations, local resignifications, and the overflowing or surplus of meaning. The latter, emerging from within the fissures of domination, expands the distance between a homogeneous order imagined and its actualization. In my view rather than an "either or" question, it is one of ethical, political, and epistemic involvement in its resignification from the macro-policy to the micro-physics and vice versa. Rather than excluding or ignoring one of the two extremes of the chain, rather than the mere antagonistic side of politics, researchers, educators, policy makers, and international agencies may contribute against inequity in visualizing, analyzing, and improving these links and their resignification, thus repositioning the two-fold face of hegemony: antagonism and articulation, inclusion and exclusion.

Notes

I am indebted to my research assistants Adriana Hernandez who collected the initial data and organized it for me to analyze and interpret it, and Sarahi López who has revised the references. I am also grateful to my colleague, Enrique

Bernal, for the information about the history of technological partnerships in Mexico. The Seminar Profundización en Analisis Politico de Discurso (1995), in Mexico has been a space for stimulating intellectual discussion for eight years. Alexis Lopez, member of this seminar, commented on the first draft. The Wednesday Group led by T. Popkewitz in Madison, Wisconsin (1999 spring term) has also been a source of intellectual exchange, and offered insightful suggestions to the first draft.

1. Saussure (1960) distinguished for analytical purposes, two interdependent constituents of the sign: *signifier*—or acoustic image—and *signified* the concept evoked by the former.
2. *Nodal point* is the sign, emblem, value, symbol, principle, or whatever signifier that operates as an apex partially stopping the flow of meaning within a discursive field; it partially condenses the sense of other signs it is articulating and to some extent, circulates throughout these signs thus making them equivalent (see Laclau, and Mouffe, 1985; also floating and empty signifiers in Laclau, 1994 and 1996).
3. *Empty signifier* is the concept to make intelligible some discursive assumptions and operations. Discursive systems are open, and rather than being articulated around an essence, they are structured around a lack of essence. This lack, this emptiness is what different discursive systems endlessly attempt to fill. The function of filling is carried out by empty signifiers, which are those signifiers capable of articulating several meanings (e.g., social demands, political values, and so on), to temporarily stop the flow of signification and thus partially stabilize a field. These signifiers are able to articulate several meanings because they reduce their links with a particular signified and thus come to represent a "universal" value. (This is an impossible and always inadequate representation because the lack of essence is not just an accident but the very condition for any system.) See also *infra* the concept of *nodal point*.
4. In Mexico 90% of private schooling is in religious hands, mainly but not only, under Catholic charge.
5. I take these signifiers as part of a chain of equivalence representing the illusion of plenitude and salvation that is regularly promoted by any party attempting to hegemonize.
6. By *deconstructive gesture* I mean not a strict, exhaustive, and rigoros Derridean reading (whatever this means); however, I will follow his logics and some basic deconstructive operations (Derrida, 1982). By a *genealogical move* I also mean my intention to follow Foucault's approach to history and the emergence of interlaced discursive series (Foucault, 1977).
7. Salvation narratives and redemption technologies are the metonymycal displacement of theological *dispositifs* (religious, sacred, or divine motives, aspirations, strategies, and other discursive resources) onto secular, Enlightenment, and rational resources (Popkewitz, 1998) by means of which in a particular epoch, a person within a community represents their ideals of plenitude, self-realization, happiness, progress, and *mutatis mutandis*—all the political and ethical ideals: progress, democracy, justice, and so on.

8. By *language game* Wittgenstein indicates convention ruled the articulation of words, acts, and objects around a purpose that constitutes them in a meaningful configuration, i.e., a discursive system in our terms, a *dispositif* in Foucauldian terms.
9. Derrida (1982) elaborates this idea of *iterability* as repetition through alteration.
10. "State" will be *used* (Wittgenstein, 1963) to mean the modern power organization form, the ensemble of institutions whereby hegemonic actions involved in a political system take place (consensus and coercion, inclusion and exclusion, articulation and antagonism, and so on); this involves the mobile frontiers between public and private. Government will be used in a rather descriptive sense of the "judiciary, legislative and executive powers," the set of political and administrative entities upon which the ruling activity has been conferred. I must distinguish this descriptive use from a conceptual use of governing as "the production of norms that separate and divide according to the available sensitivities, dispositions and awarenesses" (Popkewitz, 1998, p. 22). This conceptual use refers to the alignment of individual salvation with state planning and social goals (Popkewitz, 1998).
11. The exercise involves finding where the signifier partnership is used, with what it is associated, how it is constructed, and to what it opposes. I crisscross documents belonging to four different discursive series and indeed diverse cultural points of view and political positions. Unfortunately, this arrangement tends to decrease the visibility of the actual iterability; however it might help in reducing repetitions and thus lessening extension.
12. Condensation and displacement are the two operations involved in *overdetermination*: a mobile, multiple, and relational form of causality that fits the type of social interpretation I attempt (see Laclau and Mouffe, 1985, pp. 97–105). I will use displacement, circulation, traveling, symbolic transference as synonymous terms.
13. I have taken the Spanish *participación* as the equivalent to partnership since—in semantic terms—it remains within the same area of meaning, and—in political terms—it plays the same structuring role for domestic educational plans as it plays in international programs, and these two are the key angles I will analyze in this chapter.
14. By argumentative I mean the forms of reasoning we display to persuade our listeners of the plausibility of our world construction (Perelman and Tyteca, 1969).
15. When a lack of essence is what organizes discursive structures (social, economic, educational, etc.) we have no determinism, no algorithmic causality; instead the possibility is provided for decision.
16. Hegemonic practices, different from the commonsense idea of pure domination, imply both antagonism and articulation, inclusion and exclusion, imposition and persuasion, or if one so wishes, domination and resistance. There are thus exclusionary systems that legitimize some values and expel some other values. In other words, the most democratic system will always exclude something, and the illusion of an all-inclusive system is nothing more than a utopia too close to totalitarianism.

17. World War II involved a drastic shift in the international imaginaries thus establishing the Cold War and its geopolitical opposition between East and West (capitalism and socialism). By this time, the Mexican government—in a clear subordination to World War II—U.S. priorities and programs—produced a "rectification" of its domestic program: Cardenas's socialism was dismantled and the structuring principle was then provided by "national unity, liberalism and peace in the hemisphere."
18. By these suspicion I mean especially the intellectual movement involved among postmodern thinkers such as Lyotard, Foucault, Derrida, Laclau, and in the field of education, Popkewitz, Giroux inter alia. By postmodernity, I do not mean the abandonment, superseding, and rejection of the Enlightenment myths such as Reason, the Subject, History, and their replacement with new myths. Rather I adhere to the view of postmodernity as the undermining of the absolute character conferred on these values, and I emphasize their context-dependent and relational status. This movement, of course, was preceded by thinkers such as Nietzsche and Heidegger inter alia. (I discuss this point and its implications in Buenfil, 1997.)
19. Only when one reduces education to schooling can such a claim be conceived. If education is understood as the intellectual, cognitive, ethical, physical, and emotional process and practices whereby a biological entity becomes a person, then this claim is absurd. This is the case because the alleged monopoly the government exercised only concerned public schooling, and all the other educational agencies—e.g., family, religious temples, national rituals and festivities, sports, unions, private schools, and so on—had never been under the government dictate. On the contrary, the Catholic church has very much organized many of these other educational spaces (Buenfil and Ruiz, 1997).
20. I may seem insensitive and unfair to the differences amongst these three groups. I categorically recognize their difference. The only equivalence I sustain amongst them is their absolute denial of globalization (viewed as pure imperialism and domination), and partnership as a withdrawal of the government responsibility over public services.
21. I hold to the concept of the political as a dimension wherein there is no a priori center of politics (be it the state, class struggle, or other) and the political agent or subject (individual or collective) emerges in the precise moment of making a decision. Neither is there a center to the social. Instead any discursive centrality is precariously instituted in space and time and can therefore change. By the same token, all meanings of partnership are socially instituted and this is not a necessary reduction of government responsibility in public education. Having said this, I do not overlook the political implications of the meanings attached to this signifier in contemporary educational policies. What I want to stress is that none of these implications can stand for its ultimate truth, and consequently, there is an interstice for a decision on it; to position oneself vis-à-vis it; to pose one's own interpretation, assume our responsibility, and account for it.

22. This decision is not a fully rational, intentional, and conscious act, and is an act of free will, even less. Rather it is a decision where reason and passion, intention and spontaneity, consciousness and uncounsciousness are interwoven in politically framed contexts (see Laclau, 1990).
23. Provided that one understands ethics as different than context-dependent normativity (Butler, Laclau, and Zizek, 2000).
24. It would be difficult to ignore the unequal conditions of particular countries and the conflictive values they promote, but it would also be shortsighted to ignore the productive outcomes of this contact (further discussion in Perlmuter, 1991, Robertson, 1990, Hall, 1991, Buenfil, 2000a).
25. From a Lacanian perspective one can ask whether it is possible to have any structure at all without a promise of plenitude or fullness providing orientation to the structure. However this does not amount to saying that this plenitude is univoqual and achievable. It is rather impossible and necessary. This logic organized by aporia—an unsolvable tension between opposite arguments—enables the concept of the impossible—necessary as an intellectually productive way to think about partnership and other salvation narratives.
26. This has been my attempt in conceiving the distance and links between national educational reforms and local educational practices and asserting the relational feature constitutive of any social action. Global, national, and local are not dissociate dimensions, but rather elements—one analytically illuminates—of a sort of chain where links are fragile and changeable, a series of configurations where boundaries are mobile. In the local micro-experience one can always find traces of a national and even international macro-policy and indeed the macro-scheme can always be dislocated by a micro-event. Here one has to consider the difference between a plan and its condition of existence.

References

Secondary Sources

Bernal, E. (1998). *La educación tecnológica. Una panorámica*. México City: DIE Cinvestav (unpublished working paper).

Buenfil, B.R.N. (1990). *Politics, Hegemony and Persuasion: Education and the Mexican Revolutionary Discourse During World War I*. Unpublished doctoral dissertation, University of Essex.

Buenfil, B.R.N. (2000a). "Globalization, Education and Discourse Political Analysis. Ambiguity and Accountability in Research." *International Journal of Qualitative Studies in Education*, 13, 1–24.

Buenfil, B.R.N. (2000b). "The Mexican Revolutionary Mystique." In D.A. Howarth, A.J. Norval, and Y. Stavrakakis (Eds.), *Discourse Theory and Political Analysis. Identities, Hegemonies and Social Change* (pp. 86–99). New York: Manchester University Press.

Buenfil, B.R.N. and Ruiz, M.M. (1997). *Antagonismo y articulación en el discurso educativo. Iglesia y Gobierno* (1930–1946 and 1970–1993). México: Torres y Asociados.

Butler, J., Laclau, E., and Zizek, S. (2000). *Contingency, Hegemony, Universality*. London–New York: Verso.
Derrida, J. (1982). *Margins of Philosophy*. Brighton: Harvester Press.
Derrida, J. (1997). *El tiempo de una tesis. Desconstrucción e implicaciones conceptuales*. Barcelona: Proyecto A. Ediciones.
Dreyfus, H. and Rabinow, P. (1983). *Michel Foucault: Beyond Structuralism and Hermeneutics*. Chicago: University of Chicago Press.
Foucault, M. (1977). *Language, Counter-Memory, Practice*. Oxford: B. Blackwell.
Hall, S. (1991). "The Local and the Global: Globalization and Ethnicity." In A. King (Ed.), *Culture, Globalization and the World System* (pp. 1–17). London: Macmillan.
Laclau, E. (1990). *New Reflections on the Revolutions of Our Times*. London: Verso.
Laclau, E. (1994). "Why do Empty Signifiers Matter to Politics." In J. Weeks (Ed.), *The Lesser Evil and the Greater Good. The Theory and Politics of Social Diversity* (pp. 167–178). London: Oram Press.
Laclau, E. (1996). *Emancipation(s)*. London: Verso.
Laclau, E. and Mouffe, C. (1985). *Hegemony and Socialist Strategy*. London: Verso.
Perelman, C. and Tyteca, O. (1969). *A Treatise on Argumentation*. Notre Dame: University of Notre Dame Press.
Perlmutter, H.V. (1991). "On the Rocky Road to the First Global Civilization." *Human Relations*, 44, 897–1010.
Popkewitz, T. (1998). *Struggling for the Soul*. New York: Teachers College Press.
Popkewitz, T. (2000). "Reform as the Social Administration of the Child: Globalization of Knowledge and Power." In N. Burbules and C. Torres (Eds.), *Globalization and Education: Critical Perspectives* (pp. 157–186). New York: Routledge.
Robertson, R. (1990). "Mapping the Global Condition." In M. Featherstone (Ed.), *Global Culture* (pp. 15–30). London: Sage.
Ruiz Muñoz, M. (2000). "La participación como punto nodal en la educación de adultos." In B.R.N. Buenfil (Ed.), *En los márgenes de la educación. México a fines de milenio*. Serie: Cuadernos de Construcción Conceptual en Educación (pp. 129–144). Mexico City: Plaza y Valdes-Seminario de Análisis de Discurso Educativo.
Saussure, F. de (1960). *Course in General Linguistics*. London: P. Owen.
Wittgenstein, L. (1963). *Philosophical Investigations*. Oxford: Basil Blackwell.

Primary Sources (Official Documents)
Banco Mundial (1996). *Prioridades y estrategias para la educación: Examen del Banco Mundial*. In series *El Desarrollo en la Practica*. Washington: Banco Internacional de Reconstrucción y Fomento.
Consejo Nacional Técnico de la Educación (1992). *Acuerdo Nacional para la Modernización de la Educación Básica*. México: Dirección de Comunicación del CONALTE.

Diario Oficial de la Federación (13 Julio, 1993). *Ley General de Educación.* México, pp. 42–56.
Fondo Monetario Internacional-Asociación Internacional de Fomento (1999). *Aspectos operativos de los documentos de estrategia de lucha contra la pobreza.* http://www.imf.org/external/np/pdr/prsp/esl/poverty1.htm. (4 de febrero de 2002).
Poder Ejecutivo Federal (1989). *Plan Nacional de Desarrollo 1989–1994.* México: Poder Ejecutivo Federal.
Secretaría de Desarrollo Social (1995). *Solidaridad. Seis años de trabajo.* México: SEDESOL.
Secretaría de Educación Pública (2001). *Programa Nacional de Educación 2001–2006.* México: SEP.
UNESCO (March 1990). *World Conference on Education for All Final Report WCEFA. Meeting Basic Learning Needs.* New York: Inter Agency Commission WCEFA; Thailand.

Part II
Between the State, Civil Society, and Education

CHAPTER 3
PARTNERSHIPS IN A "COLD CLIMATE": THE CASE OF BRITAIN

Barry M. Franklin and Gary McCulloch

Probably no issue has been more central to the educational debates occurring in the United Kingdom during the last 20 or so years than that of the relative roles of the public and private sectors in the provision of education. Preoccupied with the question of how schooling could enhance national economic performance, both Conservative politicians during the 1980s and 1990s and the current New Labour government have sought ways to enlist the energies of private organizations, particularly business, in enhancing what they believe was the academic underperformance of English youth. Making in effect what was a human capital argument, policy makers from both parties claimed that the economic success of the nation hinged on the existence of a well-educated citizenry who could function in a globalized economy. Government's task, they believed, was to ensure an increase in educational standards so that British youth would possess the skills and knowledge required to improve the country's competitive economic advantage over other developed nations (Docking, 2000; Hatcher, 1998, 2001; Jenkins, 1988; Riddell, 1991; Whitty et al., 1993).

In attempting to reach this goal both the Conservatives and Labour have advocated similar educational initiatives. They have embraced market policies of competition and choice in the distribution of educational services, supported the National Curriculum, promoted accountability schemes, and advocated the establishment of public–private partnerships (Docking, 1996; Jones, 1989; McVicar, 1990; Thatcher, 1993). The focus of our attention in this chapter will be on partnerships, particularly collaboratives between the private sector and the schools. Two initiatives offer us a good picture of how this idea of partnerships has played itself out in the reform practices of both parties. Beginning in 1986, the then ruling Conservative government introduced a proposal for the creation of a number of City Technology Colleges (CTC) as a means of improving the quality of educational provision in urban schools, particularly schools serving disadvantaged students. Twelve years later, the newly elected Labour government inaugurated another partnership venture as their effort to improve educational standards in low-performing schools, its Education Action Zone (EAZ) initiative. The purpose of this essay is to examine CTCs and EAZs to see what they tell us about the efficacy of partnerships as a reform strategy. As we suggest metaphorically in the title we have given this essay, we wonder whether Britain's historical legacy of separate public and private educational spheres has provided a climate that is too cold for the flourishing of partnerships?

Our essay will bring together two lines of independent research. The first part of the article is based on Gary McCulloch's historical study of CTCs and their place in a larger history of the relationship between the state and the private sector in educational provision in Britain during the twentieth century. The second part comes from Barry Franklin's research on EAZs and more broadly on the role of partnerships in urban educational reform.[1]

The impetus for the establishment of partnerships in education and in other public services in Britain was two-fold. In part, a penchant for partnerships emerged out of the country's post–World War II economic dislocations. Neither the Keynesian policies of Labour governments nor the less interventionist actions of the Conservatives seemed able to deal with the country's pattern of slow economic growth, high unemployment, and high inflation. The demand that the International Monetary Fund imposed on Britain during the 1970s to control public expenditures in return for loans to deal with the country's economic problems led both political parties to seek alternatives to the direct state financing and operation of public services (Corry, Le Grand, and Radcliffe, 1997; Coxall and Robins, 1998).

Beyond that, however, British politicians and policy makers from both parties were seeking, to fashion a response to the economic and social problems brought about by late twentieth-century globalization. Of particular concern to them was the increased mobility of capital across national borders as a result of worldwide growth and the investment opportunities it brought. Conservatives and New Labourites alike believed that in this context government could no longer manage national economies but could only adjust policies to the demands of global markets. Toward that end they hoped to reconstruct the British state to render its markets more accessible and receptive to international capital. The strategies that they embraced shifted the state's role from directing policy to enabling it (Driver and Martell, 2001; Leys, 2001; Panitch and Leys, 2001). One vehicle for doing so was the public–private partnership. Conservative and New Labour politicians saw these collaborations as venues for private sector investment and as vehicles for bringing the financial resources and expertise of the private sector to bear on the solution of increasingly complex public problems.

Margaret Thatcher's inclination, when the Conservatives came to power in 1979, was to look to America for possible solutions to this dilemma of the state. One U.S. reform for social provision that caught her interest were the compacts that had been established between business and school systems to link the provision of jobs and private sector training opportunities for students to efforts on the part of schools to raise achievement levels. It was the visits of Conservative ministers and policy makers, first to New York in 1984 and then to Boston in 1986, that led to the formation of an array of similar business-school compacts or partnerships in London, East London, and a number of other cities throughout Britain (Richardson, 1993). Also important in this regard was the highly publicized visit of the Conservative education secretary, Kenneth Baker, to New York and other cities in September 1987, when he apparently took to praising the "can do" approach of the Americans (*Times Educational Supplement* [*TES*], 1987b; see also *TES*, 1987c).

Despite the fact that the Conservatives seemed to discover partnerships, they were an initiative that fit nicely with Prime Minister Tony Blair's efforts to remake the Labour Party in the image of a politics of the Third Way. As he began to frame his vision for the recalibration of the party into New Labour, partnerships were one of his favorite concepts for describing the relationship that he envisioned between the various public and private sectors operating in English society (Blair, 1997, 1998). Blair's interest in partnerships in fact fit nicely with his long-standing communitarian leanings that stem back to his student days at Oxford (Levitas, 1998; Rentoul, 1995).[2] Partnerships offered New Labour a strategy that would enable the party to maintain its long-standing tradition of supporting public services while at the same time allowing it to maintain its new stance of controlling public expenditures and reducing state intervention (Corry et al., 1997; Jackson, 2000).

Separate Systems

At the start, it is important to relate the idea of "partnerships" to the social and historical background of educational developments in England over the past century. In particular, the changing relationships between state and private provision during that time have created a legacy that is highly influential for recent and contemporary initiatives such as the CTCs and the EAZs (see McCulloch, 2002 and McCulloch, in press, for fuller discussions of the historical development of public and private secondary education and the relationships between them in twentieth-century England). There were several key elements in this historical experience that need to be taken into account. These include first, a tradition of "laissez-faire," which until recently led the state to take a limited role and influence in education. This meant that other agencies such as the church historically occupied a major position that they have not entirely lost. Second, was an increasingly antagonistic relationship between the state and the private sector in education that led to them being widely regarded as "two nations." The divide between them symbolized inequalities in status and social class that continued to be evident throughout the twentieth and into the twenty-first century. Third, was an increasing assertiveness on the part of the state toward education, especially from the 1970s onward. This was reflected for example in the "Great Debate" over education launched by Labour Prime Minister James Callaghan in 1976 and in the National Curriculum introduced by the Conservatives under the Education Reform Act of 1988. Some attempts were made to bring the two systems together in a working alliance, but these were obliged to operate under severe constraints that often dictated separate and even conflicting approaches to particular problems.

The "laissez-faire" dimension of this relationship had its roots in the nineteenth century, when the state was reluctant to become actively involved in the administration of a mass education system (Green, 1990). At this time there were many voluntary agencies that provided educational services for particular groups. Notably, the independent or "public" schools were able to develop as highly prestigious institutions for the social elite. The Churches—Anglican and Dissenting—were responsible for large numbers of schools around the country that were intended for working-class and middle-class children (Simon, 1960; Gardner, 1984). This situation allowed private agencies to become entrenched in the provision of education, and local and voluntary activity to be emphasized. Even when the state eventually developed a system of elementary education following the Elementary Education Act of 1870, it did so only grudgingly, and the role of private agencies was allowed to continue.

In the early twentieth century, despite the development of both an elementary and a secondary system of education under the auspices of the state, this basic reluctance to take a prominent or active role persisted. In 1927, for example, the former permanent secretary of the Board of Education, which was the central authority, could begin a discussion of its involvement with a long list of what it did not do. He pointed out, for example, that the Board did not itself directly provide, manage, or administer any schools or educational institutions except the Royal College of Art. It also had no authority over universities, university colleges, or university education. It had very little authority over endowed schools that did not receive grants. It had no authority either over schools conducted for private profit and did not make grants to them. Nor did it have authority over schools or branches of education that were in the province of other government departments, for example reformatory and industrial schools, Poor Law schools, and Army and Navy schools. The Board did not engage, pay, promote, or dismiss teachers in grant-aided schools and institutions. It did not supply, prescribe, or proscribe any textbooks for use in grant-aided schools. Moreover, the Board did not prescribe in its regulations, except in general terms, the curriculum of grant-aided schools or the methods of teaching. It also avoided interfering in particular school subjects. The Board also had no general power to interpret the Education Acts or to determine questions of law. It was not able to dissolve local education authorities; it did not provide school buildings; and it was the Ministry of Health rather than the Board of Education that sanctioned the raising of loans by local education authorities (LEA). What the Board did take responsibility for was the "superintendence" of a service "which in respect of its significance for the welfare of the community is national, and in respect of its administration and a large part of its maintenance is local" (Selby-Bigge, 1927, p. 29).

The Ministry of Education as established under the Education Act of 1944 was intended to be rather more assertive than the Board had been, but even then it was very reticent about involving itself in details of the school curriculum. These were deemed to be the responsibility of LEAs, schools, and teachers. As the Ministry declared in 1963, "in England and Wales it has long been public policy to uphold the responsibility of the schools for their own work. This is the cardinal principle of national policy in relation to the schools' curriculum and examinations.... To the maximum possible extent, each school should therefore be free to adopt a curriculum and teaching method based on its own needs and evolved by its own staff" (Ministry of Education, 1963). This in turn served to encourage a "tradition" of teacher professionalism that stressed the autonomy of teachers in the curriculum domain (Lawn, 1996; McCulloch, 2000).

The second theme that is relevant to this discussion is the polarization of state provision and private education that developed especially during the twentieth century. Private schools tended to be highly suspicious of the demands of the state, and protected their independence fiercely. On the other hand there were increasing criticisms that private schools catered to a privileged elite, and that the state was unable to provide a similar service. One possible solution, widely promoted during World War II, was to bring the independent schools into a closer association with the state system of education. Some opponents of the independent schools, such as T.C. Worsley, argued that they should be abolished (Worsley, 1940, 1941). Even some sympathizers, including Sir Cyril Norwood, a former master of Marlborough College and head of Harrow school, argued in favor of integrating them with the state system (see McCulloch, 1991). The Fleming Report of 1944, on "the public schools and the general educational system," recognized the widespread criticisms and proposals for change, and declared a need to take advantage of what it called "a unique opportunity for incorporating the public schools with their distinctive characteristics into the general system" (Board of Education, 1944, p. 4). Nevertheless, this initiative proved unable to erode the barriers between the state and independent sectors or to deliver a decisive victory to the state system. The independent schools emerged largely unscathed from the major debates of the 1940s as a separate system, comprising an alternative form of provision based on parental fees that was attractive to many because of its established social prestige. Especially from the 1960s onward, many parents with financial resources regarded the independent schools as a preferable alternative to the comprehensive secondary schools that were by then largely superseding the more academic state grammar schools.

Another aspect of the developing relationship between the state and private sectors was the increasingly active intervention of the state in primary and secondary schools. This was particularly the case after Callaghan's October 1976 speech at Ruskin College Oxford inaugurated the "Great Debate." Callaghan emphasized the importance of ensuring that the education system was accountable to parents and the public in general, in the interests of improving educational standards (Morgan, 1997). Especially under the Conservative government in the 1980s, there was a trend toward decisive intervention in the curriculum and schools, culminating in the Education Reform Act of 1988. The National Curriculum introduced under this Act was criticized even by many Conservatives as marking the "nationalization" of a service that had previously been based on local efforts. From the 1980s, too, national initiatives were produced at an accelerating pace that ensured that political demands were always prominent (see e.g., McCulloch, 2001).

Even though these strong historical threads of laissez-faire, polarization, and nationalization had tended overall to foster and reinforce a division between the state and private sectors, there were a number of experiments to try to bring them closer together or to make a more efficient and equitable use of the resources of both camps. However, these were generally stymied due to the entrenched positions of the separate systems. In the 1950s, for example, an ambitious plan to sponsor the design and equipment of science laboratories in schools, known as the "Industrial Fund," was at first intended to help bring the state and private sectors closer together. A.H. Wilson of Courtaulds Ltd., one of the main instigators of the scheme, approached the Ministry of Education to discuss how major industries could contribute to improving facilities for science education, such as paying for additions to laboratory space and equipment. He emphasized that he would not wish any industrial sponsorship to have the effect of widening the gulf that already existed between the private and state sectors (Part, 1954). The Ministry was interested in Wilson's ideas, but proved reluctant to involve state-maintained schools in such a scheme despite their lack of science facilities. The main reason for this was that local education authorities were responsible for providing equipment and facilities with the funds allocated to them. A senior government official, Toby Weaver, pointed out what he called the "familiar difficulty" that prevented private funds from being channeled into activities that were the responsibility of LEAs to maintain at a reasonable standard from public funds. He added: "Why," a private benefactor may reasonably ask, "should he subsidize public authorities in carrying out their statutory duty?" Weaver concluded that the Ministry should not ask for outside assistance toward the provision of accommodation or equipment in maintained schools (Weaver, 1957). Far from helping to bring the state and private systems together, therefore, the Industrial Fund actually reflected and maintained the divisions between them. A lack of funding to improve facilities and accommodation continued to be a problem for state-maintained schools, especially in the urban areas. One American observer in the 1960s, Richard E. Gross of Stanford University, was shocked by the inadequate finance, which led in his view to outdated facilities and poor conditions (Gross, 1965).

On the other hand, it was possible to engage in some curriculum activities that made use of private benefactors and the resources of independent schools. For example, in the early 1960s the Nuffield Foundation took the lead in sponsoring a major science teaching project that involved many teachers and schools from the independent sector (Waring, 1979; McCulloch et al., 1985). This was widely regarded, as the Director of the Foundation, Brian Young, acknowledged, as a key pump-priming initiative and an instrument for change (Young, 1965). Teachers in state-maintained schools could also seek to imitate and adapt innovations developed in independent schools (Page, 1965). Nevertheless, there remained severe constraints against practical cooperation across the sectors.

This was also evident in a further significant initiative that was developed in the 1960s, the Educational Priority Areas (EPAs). These were recommended by the Plowden Report of 1967, to help to meet the particular needs and problems of deprived social areas, especially in the inner cities. It argued that the existing formulae for allocating grants did not provide enough support for deprived areas, and concluded that "these districts need more spending on them, and government and local authorities between them must provide the funds" (Department of Education and Science [DES], 1967, p. 56). It proposed that every LEA where deprivation was found should be asked to adopt what it called "positive discrimination" within its own area, to identify and establish "educational priority schools and areas" that would qualify for favorable treatment. The Report estimated that by 1972–1973 the EPAs would add £11 million to the total current costs of maintained primary schools, and recommended that Treasury grants to LEAs should be increased to take account of this (DES, 1967, p. 65). There was no suggestion of seeking support from private sponsors. Indeed, it noted:

> Enterprising head teachers often augment their schools' allowance. Expensive equipment or additions to outdoor amenities are often paid for by money raising activities, such as jumble sales, or by the work of parent-swimming baths, pottery kilns, greenhouses, outside climbing apparatus and so on. They also help to make, maintain and repair equipment in the school. This is a powerful means of identifying them with the life of the school. But essentials must be provided by local education authorities. (DES, 1967, p. 407)

In July 1967, the government announced a special allocation of £16 million pounds for building in priority areas over the following two years. In practice, however, the DES proved highly reticent about trying to influence LEAs to implement the Plowden proposals (DES, 1972).

A national EPA program was established under the leadership of A.H. Halsey, director of the Department of Social and Administrative Studies at the University of Oxford (Silver and Silver, 1991). Halsey's report on the project, published in 1972, made some very interesting suggestions for promoting what it called a "partnership between statutory and voluntary effort" (DES, 1972, p. 184). It recommended harnessing the energies of voluntary organizations to help solve the problems of families in the EPA areas. It also pointed out several small-scale examples of cooperation in different local areas, which it argued should be encouraged. Such ideas went little further however, and in any case increasing economic difficulties in the early 1970s were already beginning to curtail the further development of expansionist programs such as this.

Overall, then, the historical development of educational provision had led to the creation of two separate systems, one public and another private, albeit with some interaction between them. This constituted an often rigid segregation that prevented or at least hindered cooperation and partnership. It provided a powerful social and historical legacy that is central to an understanding of the aims of the initiatives that were to be promoted toward the end of the century and of the problems that they faced.

City Technology Colleges

City technology colleges (CTCs) were one of the most prominent policy initiatives of the 1980s. Announced by the Education Secretary Kenneth Baker at the Conservative Party's annual conference in October 1986, they were clearly intended to mark a radical break from past policies in order to promote a new and fruitful partnership between state activity and private sponsorship. Most unlike the initiatives of the 1950s and 1960s, briefly reviewed earlier, this scheme sought actually to bypass the LEAs in establishing independent schools in urban areas. They would be supported financially from public funds, but would be run by an educational trust, and private sector sponsors would make a substantial contribution toward their costs. The CTCs, however, failed to have the impact and support that was envisaged for them, and they were quietly abandoned in the early 1990s (for fuller discussions of the CTC project see also McCulloch, 1989, 1994; Whitty et al., 1993).

The early signs of a new kind of policy were present from the beginnings of the "Great Debate" as teachers and LEAs came increasingly to be criticized for the perceived shortcomings of the education service. Such criticisms were developed with increasing vigor under the Conservative government that came to office in 1979 under Margaret Thatcher as prime minister. Comprehensive schooling was subjected to hostile scrutiny for its alleged lack of standards as well as for its supposedly liberal curriculum that failed to prepare pupils for the world of work. What became known as "privatization" began to be demanded in education as in other public services. On this view, the monopolies that provided state services should be exposed to competition and "market forces" in the interests of the consumers, who would be rewarded with more choice and diversity in educational provision, and greater accountability and quality.

A significant development toward these ends was the establishment of the Technical and Vocational Education Initiative (TVEI), launched in 1982 as a pilot scheme. This was to be administered not by the DES but by the recently created Manpower Services Commission, which was accountable to the Department of Employment. It was also to be developed "where possible" in association with LEAs, to provide technical and vocational courses for 14–18-year olds in existing schools or other educational institutions. The plan was designed to encourage greater diversity in the school curriculum and also to undermine the LEAs as the established providers. In fact, LEAs became closely involved in the project despite their early suspicions, and by 1987 it was consolidated as a national scheme with a budget of £900 million over ten years (McCulloch, 1987).

The CTCs were to be more radical still, with a design that would prevent the LEAs from taking an effective part in their development. In announcing his plan, Baker was emphatic that they would be completely free from LEA control and influence, and continued: "education can no longer be led by the producers—by the academic theorists, the administrators or even the teachers' unions. Education must be shaped by the users—by what is good for the individual child and what hopes are held by their parents" (*TES*, 1986). In this spirit, he introduced his proposal for the CTCs as a pilot network of new schools in urban areas, including the disadvantaged inner cities. Baker declared that these colleges would be for 11–18-year olds, and would not charge fees, although they would all be registered as independent schools. They would offer a curriculum with a strong emphasis on technical and scientific subjects, business studies, and design. He envisaged that about 20 CTCs would be established in the first instance, and would provide a model for many more, in, as he anticipated, a "new partnership" (*TES*, 1986). A subsequent booklet produced to celebrate this "new choice of school" emphasized that "it is desirable that LEAs and promoters of CTCs should work together for the benefit of the communities they will both serve" (DES, 1986, p. 9). Nevertheless, the terms of this partnership seemed to favor the newcomer as opposed to established interests. Baker himself insisted that he saw "no merit in protecting a monopoly" (*The Times*, 1986b). As *The Times* pointed out, "the scheme will, above all, provide the state system with competition" (*The Times*, 1986a). It was envisaged, too, that the CTCs would benefit from significant financial advantages. According to Cyril Taylor, the first chairman of the City Technology Colleges Trust and an adviser to Baker, "through the support of sponsors, as well as a more efficient use of state funding, CTCs will enjoy use of the latest hi-tech equipment, including computers, word processors, electronic mail, robots and computer-assisted design equipment" (Taylor, 1988).

The first CTC to be opened, in September 1988, was Kingshurst CTC in Solihull, and this college became in many ways the flagship for the initiative as a whole. The principal of this CTC was Valerie Bragg, a science graduate and head of Stourport-upon-Severn Comprehensive School since 1983. Kingshurst CTC was adapted for its new role on the existing premises of a former comprehensive school. The cost of adapting this site was estimated at £3.45 million. The CTC was to be partly funded by the British wing of the multinational Hanson Trust, which contributed £1 million, with a further £1 million from Lucas Industries and a number of other companies. A car manufacturer provided a car for both the principal and the director of administration and finance (Walford and Miller, 1991). The college also received equipment and laboratories from the business community, and continued to seek additional sponsorship. At the same time, it was also funded at a level similar to that of LEA schools of its size (*TES*, 1988), although it was reported that obtaining funding from the DES was a "continuing battle" and that provision was somewhat patchy. "While the CTC is well provided for in some areas, it lacks equipment in others; in only a few areas is provision actually lavish" (Walford and Miller, 1991, p. 64).

Over the following few years to the end of 1989, the government spent £46.6 million on the scheme, while sponsors had committed £12.72 million and promised a further £28.8 million. These included multinational companies like Hanson, Dixon's and ADT, alongside local figures such as Harry Djanogly, a Nottinghamshire businessman (*TES*, 1990a). Yet the scheme was already in difficulty. Fewer sponsors had come forward than had been expected, and traditional large businesses such as IBM, British Petroleum, and Imperial Chemical Industries declined to become involved. Some were perhaps concerned about what would happen to their investment in the event of a change of government, albeit Sir Gordon White, chairman of Hanson Industries, declared boldly: "I cannot think that the British population would consider disposing of a Conservative Government" (*TES*, 1987a). The private stake in the CTC program was no more than 20 percent of capital expenditure, leaving the government easily the major shareholder and spending far more money on establishing the program than had been intended (Whitty, Edwards, and Gewirtz, 1993).

Also, few suitable inner-city sites were available for the development of CTCs, with local education authorities proving predictably reluctant to sell disused schools to the CTC Trust. As a result of this, CTCs began to be opened in suburban areas, which meant also that the original mission of the scheme began to shift and became more diffuse. It was soon evident that the scheme would not reach its initial target of 20 colleges, and by 1990 the total stood at 14. In order to save the project, new ways of funding the CTCs needed to be found, so it was announced that existing schools under the aegis of LEAs could also apply to become an aided CTC, provided that they could raise a suitable amount from private sponsors. This would mean the LEAs becoming a partner in the venture, which had originally been established in direct competition with them. Taylor again radiated optimism, and predicted that there would soon be a 100 such aided CTCs (*TES*, 1990b). The government continued to express confidence in the scheme (see e.g., *TES*, 1991), in spite of a general failure on the part of the colleges to produce strong examination performance from their pupils (see e.g., *The Independent*, 1991; *TES*, 1994), but it was finally conceded in February 1994 that no further CTCs were being planned (*The Independent*, 1994).

The failure of the CTC program, despite high-level government involvement and financial support, underlined the practical difficulties in establishing an effective partnership between the public and private sector. It was conceived as an aggressively radical plan that would avoid LEA entanglements and encourage new approaches to the curriculum and learning. In practice, it still encountered strong resistance from the local education authorities and already established schools. Moreover, it was poorly thought through and confusing, with a persistent uncertainty of purpose and basic ambivalence in terms of the priorities of the colleges. What had been hailed as a "new partnership" thus became, more than anything else, a testament to the resilience of the separate systems of the public and the private spheres.

Education Action Zones

Coming to power in 1997 after almost 20 years of Conservative rule, New Labour saw educational reform as its first priority. As the government noted in its first White Paper, their goal was "education, education, education" (Department for Education and Employment [DfEE], 1997, p. 1). Key to this agenda were initiatives to ensure that schools held students accountable to high academic standards. There would be, the report goes on to say, no tolerance for failing or low-performing schools. Such schools would have to improve or be closed (DfEE, 1997). If these changes were to occur, New Labourites argued that it was necessary to reduce the power of local educational authorities over curriculum, budgets, and the hiring of staff. Doubting the ability of LEAs to raise academic standards, the government envisioned a new governance structure in which local authorities would serve as mediating institutions to enlist business, voluntary agency, and community support for improving educational performance (Letch, 2000). Partnerships were to be their vehicle for this new means of governing (Corry, Le Grand, and Radcliffe, 1997; Commission on Public Private Partnerships, 2001).

The most ambitious of these partnership ventures was the EAZ initiative. Authorized by the 1998 School Standards and Framework Act, EAZs were clusters of usually 15–25 schools located in areas of social and economic distress throughout the country. Their central purpose was to raise academic standards in low-achieving rural and urban schools thereby enhancing the social inclusion of the population. The driving force behind these zones was the establishment of partnerships among individual schools, parents, business, community organizations, and the voluntary sector. Business was to play an especially important role in both the management of these zones and their financial support. The government sought applicants from both schools and from other sectors. Local education authorities did not have to be involved but could be one of the partners or could administer a zone. By the end of Labour's first term in office there were 73 EAZs throughout Britain (DfEE, 1997, 1999b, 2000).

The EAZ initiative came with several incentives to make them attractive to schools and other potential applicants. The first 25 zones received £750,000 of government funding annually for an initial three-year period with the prospect of extending the life of the zone and its support up to five years with satisfactory performance. Each of these zones was expected to supplement this government funding by raising £250,000 in cash or in-kind annually from business or other private contributions. The 48 zones that were established subsequently received annual grants of £500,000 and matched funding up to £250,000 for each pound sterling of private sector funding that they obtained (House of Commons, 2001).

The administrative unit for a zone was its Action Forum, which served as a venue for bringing together the participating partners. Although those submitting applications for a zone developed a specific administrative structure for their forum, they typically comprised representatives from participating schools, parents, businesses, community and voluntary organizations, and the local educational authority. A Forum, in turn, appointed a project director who assumed responsibility for the day-to-day management of the zone. Once established, a Forum could assume any number of roles. It could leave the running of the schools to their existing governing bodies and focus its attention on raising standards and other overall zone targets. A Forum could, however, serve as an agent for one or more participating schools in carrying out specific zone responsibilities. Or a Forum could, if participating schools were willing to relinquish authority to it, assume responsibility for most of the functions of a zone and become the EAZ's single governing body (DfEE, 1999b).

The Case of the North Upton EAZ

The actual ability of this initiative to use partnerships as the basis of a new governing structure is, however, less clear. A government audit of the first 25 zones indicated that business contributions to these EAZs, both monetary and nonmonetary, was about half of what was expected and that most of this support was in-kind (House of Commons, 2001; Hallgarten and Watling, 2000). Research in one EAZ serving an economically depressed area within the City of London, which for the purpose of this research has been given the pseudonym of the Borough of North Upton, seems to bear this out.[3] The director of the zone commented that he had not been particularly aggressive in seeking business partners because of the financial support that the EAZ had been receiving from a large voluntary organization. He noted, however, that he was not opposed to seeking corporate contributions. "I am talking actually at this moment," he noted, "to quite a big firm who is interested in sponsoring a project in cash terms. I hope that'll be sorted out in the next two weeks." The director of another London EAZ located near North Upton *claimed* that she was quite successful in garnering business support. That support, however, came from the firm whose family trust was the principal financial supporter of this zone.

Individual schools within the North Upton EAZ did report some, albeit limited success, in attracting business support. The head teacher of one of the secondary schools in the EAZ noted that she had received

£35,000 in corporate gifts that were used to establish an ICT (information and communication technology) facility and support the school's athletic program. Another head teacher, however, pointed out that corporate monetary support was quite small. He reported receiving £400 from a local company to hold a Christmas party for his school. For him, this financial support represented "very little, miniscule amounts, £100 here, £100 there. Nothing. It's all sort of... mostly treats. Might buy the kids some sweets."

Most of the business support that schools in the zone received was in-kind. As one primary head teacher stated when asked about the type of corporate assistance she was receiving, "cash, no." The majority of support from the North Upton's business partners involved employees from these companies volunteering as tutors within individual schools. The head of one primary school noted that the Bank of England "send in their volunteers who give up their lunch time, and they work with our more able readers and develop their skills." She went on to say that they are in the process of expanding this effort and have these volunteers tutor children in mathematics. A parent at another primary school noted that employees from J.P. Morgan volunteered to work as "reading" and "number" partners to tutor both gifted and underachieving students in these areas.

Business involvement and the fear of business control of public schooling was the impetus for the opposition, particularly on the part of teachers' unions, that has plagued the EAZ program since its inception (National Union of Teachers, 2000; Socialist Teachers Alliance, 1998). North Upton was the site of the only teacher's strike in Britain that occurred surrounding this initiative. An officer of the borough's teachers association who was interviewed was particularly critical of what he saw as the EAZs "aim to attract in Tony Blair's Third Way fantasizing capital from the private sector." What ultimately worried him was that the public–private partnerships that these zones were promoting would lead to the privatization of public education. As he put it, "we don't want the curriculum brought to us by McDonalds, thank you very much indeed."

As this union official saw it, the effort of business to become involved in EAZs was part of a larger and, in his estimation, a more menacing, corporate effort. "What they're buggering around with here is for marginal profit." Their real mission, he went on to say, was to position themselves "for very big global stakes" as the providers of educational services and products in the developing world. North Upton and other zones offered them a place to "brand their product."

Equally unclear from this research in North Upton was the ability of EAZs to establish an alternative governance structure to that of the LEA. At first glance, it would appear that the role of local authorities in administering this zone was diminished. Unlike earlier EAZs that had been sponsored by their LEAs, North Upton was a joint venture of the local authority and a major voluntary sector organization, the Corporation of London. The zone's administrative structure also appeared to minimize the authority of the LEA. North Upton's Action Forum was chaired by a former head of a London financial trading company and an officer of the Corporation of London. Other members included two additional Corporation representatives; an appointee from government; and representatives from the EAZs business, community, and voluntary sector partners, from each zone school's governing body, from teachers working in zone schools, and from parents of children attending these schools. The Forum, however, met infrequently. The actual day-to-day operating authority fell to a smaller executive committee composed of the Action Forum chair, an appointee of the Corporation, a representative of one of the business partners, and three head teachers (North Upton, 1999, 2000).

Notwithstanding this administrative structure, however, the zone was to remain under the North Upton LEA's overall monitoring authority and would work collaboratively with other non-zone schools in the borough on an array of initiatives including professional development and school improvement (North Upton, 1999). The head teacher of a secondary school within the EAZ told us that she thought the LEA was playing a major role in the zone's operation, particularly in the involvement of its staff in school improvement initiatives in zone schools. North Upton's director noted that he much preferred to work with the local authority rather than with private business because in his estimation the former were more accessible and less driven by commercial concerns.

Although the role of the LEA in favor of the private sector may have been diminished in North Upton, and that is not exactly certain, the degree to which this governance structure represents an alternative to more traditional administratively dominated management schemes is less clear. The organization was hierarchical with most of the authority concentrated in a small executive board that was dominated by zone administrators. This actually seems to have been a common administrative pattern among EAZs throughout the country (Hallgarten and Watling, 2000). An officer of the North Upton Teachers Association saw this concentration of power as being deliberate. "One of the complaints that I've heard about the EAZs is you have the schools, you have the Action Forum. You have the executive committee within the Action Forum, and the whole process is one of excluding the people at the ground floor." A parent, who had served as a member of her school's governing body and was an active supporter of North Upton's EAZ, seemed to share this view. Talking about her membership on the Action Forum, she noted, "if you expressed an opinion, you were in the way really. So really, the community dropped out." The Forum, as she saw it, was primarily concerned about making "management decisions."

Like the CTC, the EAZ initiative has been short-lived. In November 2001, the New Labour government announced that the EAZ program was being disbanded and that none of the existing 73 zones would receive funding beyond the original five-year commitment. As the press and many of the individuals interviewed in North Upton and elsewhere saw it, the government was dissatisfied because the EAZs were not able to induce the hoped for financial support from the business sector and because of their inability to provide an alternative to LEA control (*TES*, 2001). Government officials took a different position. One member of EAZ team in the Department for Education and Skills noted that the lack of private sector funding was not the government's reason for abandoning the program.[4] From the beginning, he went on to say, New Labour officials did not see EAZs as permanent entities. They were, he noted, temporary initiatives that brought with them some successes that will constitute the basis for further reform. As the government saw it, he pointed out, the best strategy for the future was to integrate this program into their overall school reform strategy. Another member of the EAZ team noted that business financial support was not the most important contribution that business partnerships brought. "The cash is useful," but, she went on to say that "these businesses have specialties in more general skills such as management of staff, that sort of thing. We wanted to draw on all of that, and the cash is very nice, but it isn't really the biggest thing." But this was not an entirely convincing spin on the demise of an initiative that had once been described, by the then school standards minister, Stephen Byers, as "the test bed for the education system of the 21st century . . ., a fundamental challenge to the status quo, a real threat to those vested interests that have for too long held back our school system" (*TES*, 1998).

As it turns out EAZs will not exactly disappear. Within a year of introducing its original EAZ program, the government put forward another proposal for a number of smaller EAZs as part of its Excellence in Cities initiative for addressing low achievement in inner-city schools.

These zones, which the government has not disbanded, have similar targets to their larger counterparts but include fewer schools. And they were still to be vehicles for establishing partnerships. Yet what is different about these smaller zones is that they are to be funded and administered through local education authorities (DfEE, 1999a). What made the EAZ program an innovative reform, its potential for business financial support and for a new governance structure outside the EAZs, has been stripped away in the effort to recalibrate this initiative.

Conclusions

CTCs and EAZs may both be regarded as experiments in public–private partnerships, encouraging collaboration between business and schools, bringing together government, business, the voluntary sector and citizens, in shortchanging the boundaries between the state and the marketplace. This is not to say that they were entirely the same in their approach. Indeed, they were very different in terms of their political constituency, and the CTCs had begun as an attempt to encourage a particular kind of curriculum, which was later abandoned. They also allocated different roles to the established LEAs. No doubt in these respects the EAZs learned something from the failures of the CTCs, with key figures such as (now Sir) Cyril Taylor of the Technology Colleges Trust continuing to exert a strong strategic influence. It is also fair to see some similarities between these recent experiments in partnership, and some initiatives of the 1950s and 1960s. The Industrial Fund had also tried to break down the barriers between public and private, while the Educational Priority Areas had pioneered special provision for socially disadvantaged schools and districts. CTCs and EAZs did have something at least to build on.

Nevertheless, four features stand out here as being worthy of emphasis. The first is the underlying continuities between the CTCs and the EAZs. Despite their differences in detail and orientation, this is a prime example of how Tony Blair's Third Way could co-opt key themes in the Conservative agenda of the 1980s. The similarities of these two initiatives is a good illustration of how New Labour's education agenda perpetuated, albeit in something of a different form, the market-oriented approaches of its Conservative predecessor.

The second point to be made is about the basic discontinuities between the postwar initiatives and those of the 1980s and 1990s. The Industrial Fund of the 1950s eventually benefited only the private sector, despite its own original hopes, while the Educational Priority

Areas of the late 1960s were to be based first and foremost on public funds. Both CTCs and EAZs, on the other hand, were promoted on the grounds that they brought a mix of public and private funding to the support of education. The qualitative shift from the policies of the 1950s and 1960s to those of the 1980s–1990s also reflects a basic change in the position of the LEAs. Where once they had been fundamental and inescapable, they had become either one player among many or, especially in the 1980s, eliminated from the equation altogether.

A third feature is perhaps even more striking than these first two. It is that both the CTCs and the EAZs, for all their high-level political support and investment, found it hard to make headway. Ultimately they disappointed their supporters and justified the skepticism of their detractors. In the end, they both withered in the cold climate in which they found themselves. Undoubtedly if we are to understand this overall failure we must locate it in terms of the powerful social, cultural, and historical legacy of English education. The separate spheres of private and public provision created expectations and structures that proved highly resilient against both open assault and quiet coaxing.

Fourth, our study of CTCs and EAZs looks beyond Britain's "cold climate" for such enterprises and tells us something more generally about the prospects of partnerships as vehicles of educational reform. The providers of educational partnerships defend initiatives like the two we considered in this essay on the grounds that the state alone cannot resolve the problems that globalization has brought to contemporary society. Partnerships, they argue, are vehicles through which a reconstructed British state can enlist the resources and expertise of the private sector in resolving these dilemmas (Hatcher, 2001; Hatcher and Leblond, 2001). In the end, neither CTCs nor EAZs were able to bring together that combined effort, the so-called joined-up policy that is so popular in the New Labour lexicon (Power, 2001, p. 16).

The EAZ program as we saw in our look at North Upton was a highly divisive strategy that pitted teachers and their unions against local education authorities and the government. EAZs have in fact represented a major impetus for the reproof of New Labour on the part of many left-leaning academics and intellectuals. As these critics see it, the real education agenda of the party, despite Blair's protests to the contrary, is to achieve what the Conservatives were unable to accomplish, namely the privatization of British state schooling and the undermining of the principle of comprehensive education (see e.g., Avis, 2000; Power and Whitty, 1999; Chitty and Simon, 2001). The EAZ experience suggests that partnerships may not be the solution to the battles of left and right and public and private as has been argued but rather the perpetuator of such conflicts.

Equally troubling for our assessment of the prospects of partnerships for promoting reform are the views of the head teachers we interviewed in North Upton. They were almost uniformly supportive of this program. They felt this way primarily because EAZs brought new money into their schools, which allowed them to fund improvements that they could not have previously done. They were not particularly critical of partnerships. Yet, this approach to reform was almost beside the point to their way of thinking. If in the end, such a key group of EAZ supporters as school heads did not really care much about partnerships, one may well wonder about the long-term impact of this reform on actual educational practice.

Both CTCs and EAZs are good illustrations of the dilemma that Tyack and Cuban (1995) identify with efforts at "reinventing schooling" (p. 110). As they see it, reforms like these two partnership ventures that originated outside the schools at the behest of politicians and policy makers, often represent too much of a challenge to the taken-for-granted and accepted managerial and pedagogical practices of educational settings, what they call the "grammar of schooling," and as a consequence have a short-lived existence and minimal impact (p. 85). Resolving this issue, however, requires additional research beyond the scope of this essay. In broad terms, an interim conclusion to our research would be that the CTCs and EAZs failed to escape from inherited attitudes and structures, and failed to usher in the new world of educational partnership that their authors imagined was at hand.

In some senses, the difficulties of the CTCs and EAZs might be attributed to what Edward Thompson described as the "peculiarities of the English" (Thompson, 1978, p. 39). They were indeed played out in a fashion that was distinctive if not unique. That is, both CTCs and EAZs existed under the influence of contradictory forces, one a worldwide impetus for extending the private sector's role over educational provision and the other, the quintessential British tradition of separate public and private educational spheres. Nevertheless, they do offer broader lessons for the conceptualizing and advocacy of partnership arrangements. In particular, partnerships need to be approached and assessed in their historical context, that is, in relation to the social, political, and cultural history of the country, area, or case involved. Many of the advocates of the CTCs and EAZs evinced an ahistorical, future-oriented outlook, or as Baker termed it a "can-do" approach that ignored or sought to override the historical issues that would do so much to undermine their efforts (*TES*, 1987b). Greater awareness of the historical dimension should give a clearer sense of the problems to be faced, the nature of the partnerships that can be contemplated and attempted, and the prospects for their ultimate success.

Notes

1. Barry Franklin's research was supported by a Faculty Development Grant from the Office of Research at the University of Michigan-Flint and a Rackham Faculty Grant from the Horace Rackham Graduate School at the University of Michigan-Ann Arbor.
2. Blair is not alone among New Labourites in supporting partnerships as a reform initiative. Michael Barber who is currently a key Blair adviser on education and was formerly Head of the Standards and Effectiveness Unit within the Department for Education and Employment (DfEE) was promoting partnerships as an important strategy for improving urban schools since at least the early 1990s when he was Director of the Centre for Successful schools at Keele University. See Barber and Johnson, 1996.
3. North Upton is a pseudonym for the real community in which this phase of the research for this essay was conducted by Barry Franklin during January and February 2001, November 2001, and March 2002. During those visits to the zone, he interviewed members of the government's EAZ team, the director of the zone and directors of two nearby zones, head teachers of five of the 16 schools within the zone, a parent and governor of one of the schools within the zone, a business member of the Action Forum and Executive Board of the zone, and two community activists working in agencies within the zone. Those participating in these interviews were guaranteed anonymity and are thus not identified in the study.
4. Following Blair's election to a second term in 2001, the name of the DfEE was changed to the Department for Education and Skills (DfES).

References

Avis, J. (2000). "The Forces of Conservatism: New Labour, the Third Way, Reflexive Modernization and Social Justice." *Education and Social Justice*, 3, 31–36.

Barber, M. and Johnson M. (1996). "Collaboration for School Improvement: The Power of Partnerships." In M. Barber and R. Dunn (Eds.), *Raising Educational Standards in the Inner Cities: Practical Initiatives in Action* (pp. 129–142). London: Cassell.

Blair, T. (1997). *New Britain*. Boulder: Westview Press.

Blair, T. (1998). *The Third Way: New Politics for the New Century*. London: The Fabian Society.

Board of Education (1944). *The Public Schools and the General Educational System*. London: Her Majesty's Stationery Office (HMSO).

Chitty, C. and Simon, B. (Eds.) (2001). "*Promoting Comprehensive Education in the 21st Century*." Stoke on Trent: Trentham Books.

Commission on Public Private Partnerships (2001). *Building Better Partnerships. The Final Report of the Commission on Public Private Partnerships.* London: Institute for Public Policy Research.
Corry, D., Le Grand, J., and Radcliffe, R. (1997). *Public–Private Partnerships. A Marriage of Convenience or a Permanent Commitment.* London: Institute for Public Policy Research.
Coxall, B. and Robyns, L. (1998). *British Politics since the War.* London: Macmillan Press, Ltd.
Department for Education and Employment (DfEE) (1997). *Excellence in Schools.* London: DfEE.
DfEE (1999a). *Excellence in Cities.* London: DfEE.
DfEE (1999b). *Meet the Challenge: Education Action Zones.* London: DfEE.
DfEE (2000). *Education Action Zones—A Real World Success Story.* London: DfEE.
Department of Education and Science (DES) (1967). *Children and Their Primary Schools.* London: HMSO.
DES (1972). *Educational Priority: Volume 1.* EPA Problems and Policies. London: HMSO.
DES (1986). *A New Choice of School: City Technology Colleges.* London: HMSO.
Docking, J. (1996). "School Choice." In J. Docking (Ed.), *National School Policy: Major Issues in Education Policy for School in England and Wales 1999 onwards.* London: David Fulton Publishers.
Docking, J. (2000). "What is the problem?" In J. Docking (Ed.), *New Labour's Policies for Schools: Raising the Standard?* (pp. 3–20). London: David Fulton Publishers.
Driver, S. and Martell, L. (2001). "Left, Right and the Third Way." In A. Giddens (Ed.), *The Global Third Way Debate* (pp. 36–49). Cambridge: Polity Press.
Gardner, P. (1984). *The Lost Elementary Schools of Victorian England. The People's Education.* London: Croom Helm.
Green, A. (1990). *Education and State Formation: The Rise of Education Systems in England, France and the USA.* New York: St. Martins Press.
Gross, R.E. (1965). "British Secondary Education—an Appraisal." In R.E. Gross (Ed.), *British Secondary Education: Overview and Appraisal* (pp. 547–577). London: Oxford University Press.
Hallgarten, J. and Watling, R. (2000). "Zones of Contention." In R. Lissauer and P. Robinson (Eds.), *A Learning Process: Public Private Partnerships in Education* (pp. 22–43). London: Institute for Public Policy Research.
Hatcher, R. (1998). "Labour, Official School Improvements and Equality." *Journal of Education Policy,* 13, 485–499.
Hatcher, R. (2001). "Getting Down to Business: Schooling in the Globalized Economy." *Education and Social Justice,* 3, 45–59.
Hatcher, R. and Leblond, D. (2001). "Education Action Zones and Zones d'Education Prioritaires." In S. Riddell and L. Tett (Eds.), *Education, Social Justice and Inter-agency Working: Joinedup or Fractured Policy?* (pp. 29–57). London: Routledge.

House of Commons (2001). *Education Action Zones: Meeting the Challenge-the Lesson Identified from Auditing the First 25 Zones*. Report by the Comptroller and Auditor General. London: The Stationery Office.

Jackson, P. (2000). "Choice, Diversity and Partnerships." In J. Docking (Ed.), *New Labour's Policies for Schools: Raising the Standard?* (pp. 177–190). London: David Fulton Publishers.

Jenkins, P. (1988). *Mrs. Thatcher's Revolution: The Ending of the Socialist Era*. Cambridge: Harvard University Press.

Jones, K. (1989). *Right Turn: The Conservative Revolution in Education*. London: Hutchinson.

Lawn, M. (1996). *Modern Times? Work, Professionalism and Citizenship in Teaching*. London: Falmer Press.

Letch, R. (2000). "The Role of Local Education Authorities." In J. Docking (Ed.), *New Labour's Policies for Schools: Raising the Standard?* (pp. 158–176). London: David Fulton Publishers.

Levitas, R. (1998). *The Inclusive Society? Social Exclusion and New Labour*. Macmillan Press Ltd.

Leys, C. (2001). *Market Driven Politics: Neoliberal Democracy and the Public Interest*. London: Verso.

McCulloch, G. (1987). "History and Policy: The Politics of the TVEI." In D. Gleeson (Ed.), *TVEI and Secondary Education* (pp. 13–37). Buckingham: Open University Press.

McCulloch, G. (1989). "City Technology Colleges: An Old Choice of School?" *British Journal of Educational Studies*, 37, 30–43.

McCulloch, G. (1991). *Philosophers and Kings: Education for Leadership in Modern England*. Cambridge: Cambridge University Press.

McCulloch, G. (1994). *Technical Fix: City Technology Colleges*. Leeds: University of Leeds.

McCulloch, G. (2000). "The Politics of the Secret Garden: Teachers and the School Curriculum in England and Wales." In D. Day, A. Fernandez, T. Hauge, and J. Moller (Eds.), *The Life and Work of Teachers: International Perspectives* (pp. 26–37). London: Routledge.

McCulloch, G. (2001). "The Reinvention of Teacher Professionalism." In R. Phillips and J. Furlong (Eds.), *Education, Reform and the State: Politics, Policy, and Practice, 1976–2001* (pp. 103–117). London: Routledge.

McCulloch, G. (2002). "Secondary Education." In R. Aldrich (Ed.), *A Century of Education* (pp. 31–53).

McCulloch, G. (in press). "From Incorporation to Privatization: Public and Private Secondary Education in Twentieth Century England." In R. Aldrich (Ed.), *Private and Public: Studies in the History of Knowledge and Education*. London: Woburn.

McCulloch, G., Jenkins, E.W., and Layton, D. (1985). *Technological Revolution? The Politics of School Science and Technology in England and Wales since 1945*. London: Falmer.

McVicar, M. (1990). "Education Policy: Education as a Business." In S. Savage and L. Robbins (Eds.), *Public Policy under Thatcher* (pp. 131–144). London: Macmillan Press Ltd.

Ministry of Education (September 1963). "Memorandum to the Working Party on Curriculum and Examinations: The Outlines of the Problem." *Ministry of Education Papers*. Public Record Office, ED 147/814.
Morgan, K. (1997). *Callaghan: A Life*. Oxford: Oxford University Press.
National Union of Teachers (2000). *An Evaluation of Teachers in Education Action Zones: Executive Summary*.
North Upton (pseudo.) (1999). *Education Action Zone Application*.
North Upton (pseudo.) (2000). *EAZ Action Plan*.
Page, G.T. (1965). *Engineering among the Schools*. London: Institute of Mechanical Engineers.
Panitch, L. and Leys, C. (2001). *The End of Parliamentary Socialism: From New Left to New Labour* (2nd ed.). London: Verso.
Part, A. (July 30, 1954). Note of an Interview with A.H. Wilson. *Ministry of Education Papers*. Kew, ED 147/211.
Power, S. (2001). "Joined-Up Thinking? Inter-Agency Partnerships in Education Action Zones." In S. Riddell and L. Tett (Eds.), *Education, Social Justice and Inter-Agency Working: Joined-Up or Fractured Policy?* (pp. 14–28). London: Routledge.
Power, S. and Whitty, G. (1999). "New Labour's Education Policy: First, Second or Third Way?" *Journal of Education Policy*, 14, 535–546.
Rentoul, J. (1995). *Tony Blair*. London: Little, Brown and Company.
Richardson, W. (1993). "Employers as an Instrument of School Reform? Education—Business 'Compacts' in Britain and America." In D. Finegold, L. McFarland, and W. Richardson (Eds.), *Something Borrowed, Something Learned? The Transatlantic Market in Education and Training Reform* (pp. 171–192). Washington, D.C.: The Brookings Institution.
Riddell, P. (1991). *The Thatcher Era and Its Legacy*. Oxford: Blackwell.
Selby-Bigge, L.A. (1927). *The Board of Education*. London: Putnam and Sons.
Silver, H. and Silver, P. (1991). *An Educational War on Poverty: American and British Policy-Making, 1960–1980*. Cambridge: Cambridge University Press.
Simon, B. (1960). *Studies in the History of Education*. London: Lawrence and Wishart.
Socialist Teachers Alliance (1998). *Trojan Horse—Education Action Zones: The Case Against Privatization of Education* (2nd ed.). Walthamston: Socialist Teachers Alliance.
Taylor, C. (January 22, 1988). "Climbing Towards a Skillful Revolution." *Times Educational Supplement (TES)*.
Thatcher, M. (1993). *The Downing Street Years*. London: HarperCollins.
The Independent (October 18, 1991). "CTC's technology teaching 'less than satisfactory.'"
The Independent (February 19, 1994). "City College Scheme Ends."
The Times (October 8, 1986a). "Mr. Baker's Parent Power."
The Times (October 15, 1986b). "Storm of Criticism for New Colleges."
(TES) (October 10, 1986). "Baker's Course to Put Governors at the Helm."
TES (February 27, 1987a). "Baker Buoyant on Mid-Atlantic Float."
TES (September 25, 1987b). "Baker Thinks We 'Can Do' Too."
TES (October, 2 1987c). "No Strikes—Except to Strike it Rich."

TES (September 16, 1988). "Newcomer Worries the Neighbors."
TES (June 1, 1990a). "Baker Offspring Dear to Rear."
TES (June 1, 1990b). "Changing Tack to Stay on Course."
TES (July 5, 1991). "Major Revives CTCs Project."
TES (November 25, 1994). "High-Cost CTCs Fall Short of Excellence."
TES (26 June, 1998). "You are Entering the Try-Out Zones."
TES (November 16, 2001). "Action Zones to be Scrapped."
Thompson, E.P. (1978). "The Peculiarities of the English." In: E.P. Thompson, *The Poverty Of Theory, and Other Essays* (pp. 35–91). London: Merlin Press.
Tyack, D. and Cuban, L. (1995). *Tinkering Toward Utopia*. Harvard: Harvard University Press.
Walford, G. and Miller, H. (1991). *City Technology Colleges*. Buckingham: Open University Press.
Waring, M. (1979). *Social Pressures and Curriculum Innovation: A Study of the Nuffield Foundation Science Teaching Project*. London: Methuen.
Weaver, T. (November 19, 1957). Memo to R.N. Heaton. *Ministry of Education Papers*. Kew, ED 147/794.
Whitty, G., Edwards, T., and Gewirtz, S. (1993). *Specialisation and Choice in Urban Education: The City Technology College Experiment*. London: Routledge.
Worsley, T.C. (1940). *Barbarians and Philistines: Democracy and the Public Schools*. London: Robert Hale and co.
Worsley, T.C. (1941). *The End of the "old school tie."* London: Secker and Warburg.
Young, B. (November 26, 1965). Letter to E.H. Goddard (headmasters' conference). *Head Masters' Association Papers*, University of Warwick Modern Records Centre, 58/3/4.

Chapter 4
Education Action Zones: Model Partnerships?[1]

*Marny Dickson, Sharon Gewirtz, David Halpin,
Sally Power, and Geoff Whitty*

The English Education Action Zone (EAZ) policy is one of a number of area-based initiatives[2] introduced by the first New Labour administration (1997–2001) which aims to develop integrated solutions to complex social problems within regions of social disadvantage (Glendinning et al., 2002). Intended to develop innovative local solutions to social exclusion and educational underachievement, at the time of the policy's launch in January 1998, the EAZ initiative was described by a key government minister as a "forerunner for the future delivery of public services in the next century" (Byers, cited by Rafferty, 1998, p. 4). The concept of partnership is at the heart of the EAZ policy—partnership between different public sector providers and between private, voluntary, public and community sectors—and embodies two core New Labour beliefs about the value of partnerships. The first is that neither the state nor the market acting on their own can provide an adequately responsive and fair way of organizing welfare services. The second is that "joined-up problems" require "joined-up solutions." In other words, the view is that different government departments and different welfare sectors need to work together in order to ensure that individuals receive services that are efficient, coordinated, and integrated rather than inefficient, fragmented, and contradictory (Cabinet Office, 1999). In what was sometimes presented as a "win win" situation, it was claimed that EAZs would combine the strengths of public, private, voluntary, and community organizations to create additional benefits all round. Some advocates of the policy went as far as to argue that the emphasis on the involvement of community partners in running the zones provided the possibility of a more collaborative and inclusive politics of education. Others have been more skeptical about the benefits of private sector involvement in the provision of state schooling and the potential for effective collaborative working within a marketized context, and have expressed concerns about possible power imbalances and clashes of interest amongst the various "partners."

In this chapter, we draw on evidence from an ongoing Economic and Social Research Council project,[3] which has examined the origins, implementation, and effects of EAZ policy, to subject to empirical scrutiny some of the claims and counterclaims made about the kinds of partnerships EAZs were meant to be piloting. In particular, we consider whether EAZs have indeed managed to facilitate collaborative and inclusive modes of educational governance and the extent to which private sector involvement in EAZs has helped to revitalize schooling in these areas. We also note the paucity of multiagency and interschool partnership initiatives and identify some of the factors that have inhibited effective "joined-up" working. Although the EAZ initiative is a specifically English policy, we suggest that there are lessons to be learnt from the policy about the functioning of partnerships that are relevant to other national contexts where partnership working is being advocated as a way forward for welfare delivery. We begin though by briefly identifying the main elements of the initiative, the different kinds of partnership envisaged within EAZ policy documents, and competing claims about the purpose and significance of EAZ partnerships.

Understanding the EAZ Partnership Agenda

Allocated via a process of competitive bidding, 25 "first-round" EAZs were introduced between September 1998 and January 1999, followed by a further 48 "second-round" zones in the period September 1999–April 2000. The assumption in the Department for Education and Employment (DfEE)[4] guidance was that a small number of main "partners" would put forward each EAZ proposal. This would "typically include one or more representatives of the following: a parent group, a local education authority (LEA), a training and enterprise council (TEC), a business or a community or voluntary organisation" as well as at least one representative from the participating schools. Other "organisations" were "likely to be actively involved once a zone is running" (DfEE, 1998a, p. 1). It was hoped that EAZ partnerships would "draw in local and national agencies and charities involved in, for example, health care, social care and crime prevention" (DfEE, 1997, p. 4) and that EAZs would "link up" with Health and Employment Zones and projects funded by the Single Regeneration Budget.

A typical EAZ consists of around 20 schools (usually two or three secondary schools plus their feeder primaries). Although EAZs are managed on a day-to-day basis by an appointed director, each EAZ is formally governed by an Education Action Forum (EAF), a statutory body set up through powers established by the 1998 School Standards and Framework Act. The EAFs are required to include one appointee from each of the governing bodies of participating schools (with the caveat that any of the governing bodies can decide they do not wish to be represented) and one or two members appointed by the Secretary of State. In addition, each EAF may add to its membership parents, students, and representatives of the local "business and social community." EAFs each have a statutory responsibility for formulating, implementing, and monitoring a detailed local action plan. To support them in this task, EAZs receive government funding of up to £750,000 per annum for three–five years, which they are expected to supplement with £250,000 per annum sponsorship in cash or "kind" from the private and/or voluntary sector. In order to implement their action plans, EAZs are encouraged to experiment with new forms of school governance, to disapply the national curriculum, and to deviate from national agreements on teachers' pay and conditions. However, despite this emphasis on local innovation, central government has retained tight control over the policy outputs that EAZs are expected to achieve. EAZ schools are not allowed to opt out of standardized national assessment tests and they are expected to set targets for pupil performance, attendance, and exclusions that are more demanding than those of non-zone schools.

The promotion of partnerships as a means of addressing the so-called wicked issues of ill health, educational underachievement, community safety, and social exclusion is not entirely new. For example, previous Conservative governments established Training and Enterprise Councils and Urban Development Corporations in order to bring public and private sector organizations together to work on urban regeneration projects. Local policing, drug prevention, and community safety initiatives have a similarly long history (Newman, 2002). Nonetheless, it can be argued that joint-working arrangements under the Conservatives were primarily a *by-product* of their policies to contract out public services and encourage the proliferation of organizations competing as service providers. As provision became more and more fragmented, networking between organizations became increasingly important (Alexander, 1991; Clarence and Painter, 1998). In contrast, the New Labour government promotes partnerships as *intentional* outcomes of central government policy initiatives (Clarence and Painter, 1998), and

Table 4.1 Proposed EAZ partnerships

Type of partnership	Rationale
Public–Private	• Utilize private sector managerial expertise • Pilot new contracts for the provision of services • Promote private sector investment in public schools • Support "work-related" learning
Cross-welfare/multiagency	• Coordinate local welfare services • Develop innovative new services
Interschool	• Counter ill effects of quasi-market competition; promote local sharing of good practice • Improve transition between phases
Parent–school	• Encourage parental support for and involvement in their children's education • Promote lifelong education
Community–EAF	• Foster civic engagement • Provide relevant services that meet the needs of local people

has adopted a more collaborative discourse. Key documents emanating from several government departments have placed an explicit and recurring focus on partnership as both a form of governance and a means of delivering a range of new "joined-up," crosscutting policies (e.g., DH, 1998; Home Office, 1998; Cabinet Office, 1999; DETR, 2000).

However, the notion of partnership is often used in rather a loose way so that its meaning is not always made clear. As Rob Atkinson (1999, p. 63) has written, "government has been unwilling to spell out exactly what partnership means, other than expressing hopes that greater co-ordination and synergy will focus minds and maximise resources." Within EAZ policy texts, the term partnership appears to encompass a range of different and quite distinct relationships (in table 4.1).

As table 4.1 indicates, the term "partnership" lacks specificity. This vagueness, however, has some concrete political benefits. In the case of partnership, the term never seems to be used in a negative sense, and because it is associated with highly regarded qualities such as inclusiveness, cooperation, and collaboration, it is difficult to criticize without appearing to favor the apparent alternatives—conflict, fragmentation, and hierarchy.

The range of relationships EAZs are expected to foster and the varied rationale for the involvement of the different players has led to competing claims about the purpose and significance of the EAZ partnership agenda. The majority of media—and academic—attention has focused on the role of the private sector partners. In the initial EAZ bidding guidance, the DfEE expressed a particular desire to support zones that were "primarily business led" (DfEE, 1998a, p. 2), thus representing the first occasion in which businesses have been encouraged to take a lead in managing *groups* of English schools. This guidance also allowed the governing bodies of zone schools to cede all of their powers to their Forum, a development that raised the prospect of a business-led Forum contracting to provide educational services in local schools and, more radically, gaining control over delegated school budgets. It was this expectation that businesses would play a lead role in the running of EAZs, coupled with a reduced role for individual school governing bodies, that led a number of observers to suggest that the policy represents part of a wider effort by government to establish new (privatized) forms of educational governance with reduced powers for LEAs and teachers (Hatcher, 1998; Jones and Bird, 2000). Predictably, the notion that EAZ public–private partnerships would "open the door for the corporate profit-making agenda" (Hatcher, 1998, p. 14) was greeted with some concern by those on the political Left, who were particularly worried about the possible commercialization of the curriculum (STA, 1998; Wilby, 1998), and welcomed by those on the right (Tooley, 1998). In New Labour texts in general, and EAZ policy texts in particular, the advantages of public–private partnerships tend not to be explicated in any detail if at all. That public–private cooperation is a good thing is usually presented as a taken-for-granted fact and therefore not something that needs to be explained or justified. However, the often unstated implication is that the "entrepreneurial zeal" (Blair, 1998) of the private sector will help to revitalize education in disadvantaged areas where the public sector is viewed as having consistently failed to offer services of adequate quality (Barber, 1998a; cited in Jones and Bird, 2000).

Alongside public–private partnerships, EAZs were also expected to pilot new cross-welfare, multiagency partnerships. This reflects a wider New Labour concern to promote "joined-up government" (Cabinet Office, 1999). This concern is based, in part, on a recognition that the fragmented and bureaucratic nature of existing welfare provision has contributed to the marginalization of welfare recipients and that the interconnected nature of social exclusion has rendered single-issue responses inadequate (Power, 2001). The EAZ policy was one of a number of New Labour initiatives designed to promote joined-up government to tackle social exclusion. At central government level these include the establishment of the Social Exclusion Unit, the Performance and Innovation Unit, the Regional Co-ordination Unit, the Neighbourhood Renewal Unit, and the Urban Policy Unit—all of which have a brief to improve the coordination of policy responses to the multiple interrelated causes and consequences of poverty and social disadvantage. However, a number of commentators have drawn attention to the long history of government attempts to promote greater integration both between government departments and between welfare agencies and to the practical difficulties of dissolving deeply entrenched interdepartmental and interagency boundaries (6, Perri, 1997; Wilkinson and Appelbee, 1999; DETR, 2000; Powell and Glendinning, 2002).

A third kind of partnership EAZs were meant to pilot were new forms of collaboration between schools and between schools and colleges. In particular, zone schools were expected to find ways of sharing good practice, for example, by employing "Advanced Skills Teachers" to work across schools or moving staff between special schools, pupil referral units, and mainstream schools, and to develop innovative strategies for promoting greater continuity between the primary and lower secondary school curriculum (DfEE, 1999). However, alongside the encouragement of such collaborative strategies, New Labour has also consolidated and extended some of the marketization strategies introduced by their Conservative predecessors, so that schools are still in competition with each other for student recruitment. Some early commentators drew attention to the tension between the government's expectation of interschool collaboration and its perpetuation of local education markets, noting, "to innovate in collaborative ways with other (zone) schools would have to overcome the division and conflicts which markets can produce at a local level" (Gewirtz, 1999, p. 155).

Finally, the EAZ policy was meant to facilitate more collaborative relationships between schools and their local communities, including parents. Some commentators went as far as to claim that the policy could foster new forms of democratic engagement and provide real opportunities to develop a more open and inclusive politics of education (Barber, 1998b; Halpin, 1999). However, the diversity of the proposed EAZ partnerships prompted others to be more skeptical about "the extent to which the various players are 'partners' in any meaningful way" (Power and Whitty, 1999, p. 544). Such commentators speculated that whilst parental "involvement" may be secured, it was unlikely to be on the terms of those whom the policy targeted as a "problem" (Power and Gewirtz, 2001). Some of the more skeptical commentators also noted that, if the term partnership serves to disguise power imbalances between the different "partners," it also conceals the extent to which the interests of these "partners" are likely to clash (Hatcher, 1998; Gewirtz, 1999; Jones and Bird, 2000).

EAZ Partnerships in Practice

In the remainder of this chapter we draw on research from our Economic and Social Research Council study in order to consider how EAZ partnerships operated in practice. In doing so we will assess some of the claims and counterclaims summarized earlier. It would not be possible in the space available here to explore in depth every aspect of partnership working identified in table 4.1 or to evaluate every claim and counterclaim made about the policy. Instead we focus on three areas of inquiry. We begin by examining the operation of EAZ partnerships at a strategic level. We analyze the assumptions about the membership and purposes of the EAFs that were embedded in the successful applications for first-round zone status and draw on observational and interview research to explore the decision-making practices of EAFs. Our particular focus here is the extent to which EAFs are fostering new forms of civic engagement and greater community involvement in local education decision-making. We then consider the role of private sector partners in terms of the extent and nature of their investment and involvement in participating schools, in order to reflect on the consequences of these "new" partnerships for their intended beneficiaries—schools and their students in areas of disadvantage. Here our focus is on the extent to which private sector involvement can be said to contribute to a revitalization of education provision in the areas studied. Finally, we draw attention to the limited extent of multiagency and interschool collaboration in the zones and consider some of the barriers to "joined-up working" that are being experienced at the school level as a consequence of tensions inherent within the EAZ policy and the wider policy context in which EAZs are embedded.

Community Partnerships? Strategic Decision-Making and Internal Democracy

Advocates of EAFs argued that these new partnerships would promote an open and inclusive form of educational governance that would "empower" local people. In this section we consider the extent to which EAFs have succeeded in meeting this high expectation by focusing on three key issues: representation (which partners are included or excluded on EAFs?); function (the purpose of EAF meetings); and ultimate authority (which partners are most influential?). To answer these questions, we draw on documentary analysis of the first 25 successful applications for zone status as well as observations of EAF meetings and interviews with Forum members in two zones (Tolside and Nortown—both pseudonyms).

In relation to representation, our analysis of successful EAZ applications (discussed in more detail in Dickson et al., 2001) illustrates that the composition of proposed EAFs was skewed in favor of certain types of partners. In particular, the emphasis placed on individual institutional representation ensured that representatives of the participating schools were numerically dominant on the majority of EAFs. While headteachers' voices were well represented in EAZ bids, other constituents, particularly parents and community groups, struggled for recognition. Less than a third of the first 25 zones, for example, stated in their original bids that their Forums would have either direct parent or student involvement; just nine mentioned teacher representatives and only three highlighted the role of teaching trade union officials. Strikingly, only a handful of bids made explicit commitments to involve the wider public and only two described the EAZs role as one of increasing opportunities for local democracy.

In general, named business partners were well represented in EAZ applications and bids often made explicit commitments to promoting "joined-up working" between different welfare sectors. However, as Jones and Bird (2000, p. 505) argue, EAFs "are not arenas where a broad range of interests . . . encounter each other." It was unusual in EAZ applications to find representation of welfare providers *other* than those in education, and the private and voluntary sector organizations that were represented typically concentrate on areas involving the provision of educational services or work-related training. As a result, although EAZs acknowledge the impact of social and economic factors on educational attainment, it is predominantly educational solutions that they propose (Power, 2001).

It is of course precipitate to judge the nature or operation of EAFs on the basis of hastily put together EAZ applications. In the course of our research we have attended 37 EAF and executive board meetings in Nortown, a first-round zone, and Tolside, a second-round zone over a three-year period. These observational data have been supplemented by interviews with 16 Forum members across the two zones. The populations served by the two EAZs, the composition of their Forums, meeting cycles and relationships with other committees varies enormously. Nortown is an inner-city zone serving a culturally diverse population characterized by high levels of socioeconomic disadvantage. By way of contrast, Tolside covers a much less culturally diverse town population, which although educationally "under-performing," compares rather favorably to Nortown on the relevant socioeconomic indicators. However, like many other nonurban EAZs, Tolside faces particular difficulties in its attempts to meet aspects of the EAZ partnership agenda due to a lack of large local employers. Consequently, while Nortown has several national and multinational businesses named as EAZ partners, Tolside has relied primarily on small local companies. Despite these differences, there are strong similarities in the nature and operation of the Nortown and Tolside EAFs.

While *some* EAZ partners and other interests have formal positions on the two EAFs, their composition does not reflect the character of either of the zone communities. In numerical terms, headteachers and business partners are well represented (even if in Tolside the latter rarely attend), while teachers are proportionately underrepresented. There is very limited involvement of either parents or local voluntary and community group representatives on the Forums; moreover, the related executive boards and working committees (each consisting of around ten members in total) have even fewer, if any, classroom teachers, community, or voluntary group representatives. Both EAZs have found it difficult to tap into existing voluntary or community groups and organizations, let alone to develop innovative forms of engaging previously excluded groups. In Nortown it is particularly noticeable that community representation is heavily skewed away from the character of the district in which the zone is located—a highly multiethnic population living in poor housing in an area of high unemployment. Rather than reflecting these characteristics, as the DfEE bidding guidance instructed, the majority of EAF members are white, male, and in professional or managerial positions. In Tolside, the gender composition is more even (due to the high numbers of primary headteachers represented), but no non-white Forum members have been present at any of the observed meetings.

Instability of Forum membership and low attendance levels at meetings compound the concerns we have identified in relation to EAF composition. Tolside and Nortown have both found it difficult to replace Forum members (particularly business and community members) who have moved jobs, been promoted, or simply not attended meetings. As a consequence, attendance at EAF meetings in the two zones frequently runs at less than 50 percent and numerous meetings have been inquorate. Moreover, despite having a Forum that formally consists of about

50 members, Tolside has *never* had more than three partners from outside the publicly funded education sector attend any observed Forum meeting.

The character of Forum meetings raise further questions about the extent to which EAZs are fostering a new more participatory politics of education. Our analysis of the first 25 EAZ bids shows that the intended size of individual EAFs ranged from around 20 to over 50. It is not surprising, therefore, that the majority of aspiring EAZs (including both Nortown and Tolside) intended to establish some form of executive group that would meet fairly frequently (often monthly) to make and act upon decisions, thereby reducing the role of the EAF to one of providing "ultimate" approval. Consequently, despite the formal responsibility of EAFs to provide their zone with "strategic direction" (DfEE, 1999, p. 7), in practice it appears that both strategic and operational decisions are often made elsewhere. In both Nortown and Tolside, EAF meetings serve primarily as a venue where information is passed on to the various constituencies represented. Decisions and overall progress are typically *reported* rather than discussed or debated—and there is little time for members to introduce new issues. As the director of Tolside acknowledged, "[the Forum] simply confirms policy." Forum members in both zones reinforced this perception:

> The one time I've been to it [a Forum meeting] it didn't seem to make any decisions. It was more a body that received reports from the lead managers in the EAZ and unless there were any glaring matters of probity or whatever that we should be deciding about, then it was a report receiving Forum, rather than a steering, leading, driving type Forum. (Voluntary sector partner, Tolside)

> The Forum doesn't manage anything, if we're really honest, it's a way of bringing all the people together so at least there's an information sharing . . . at the moment we neither manage nor influence strategy, we listen to a reporting team . . . the nature of the way they [the EAZ director and LEA] work will be bringing things to the Forum at best for ratification and development rather than decisions and initiation. (Business partner, Nortown)

While it is always likely that attendance will be highest during the early stages of an initiative, the interviews we have conducted with EAF members suggest that the decline in attendance experienced in both Tolside and Nortown reflects a decreasing sense of engagement with these Forums. It appears that some EAF members began to feel rather redundant when they discovered that in practice rather than making decisions they were primarily confirming decisions that had been made elsewhere. In some cases this led to a considerable level of disillusionment and disaffection. Two "lapsed" Forum members recall:

> I felt that everything that I'd said had fallen on deaf ears, we hadn't achieved anything, hadn't gone anywhere really . . . The other community people that are on the Forum, they're not anymore. They have all gone. (Community partner, Nortown)

> I honestly don't see what the role of the parent rep is. I really don't know what they're there for . . . It turned out to be in the end, as I saw it, no more than a rubber-stamping role to be truthful with you . . . I really just felt that that's all I was there for, just another voice, or just another body, so to speak, and they could say, oh, well, the parents were represented. (Parent representative, Tolside)

Perhaps unsurprisingly, it is the members who have the most knowledge of and involvement in managing the zone on a day-to-day level that dominate EAF meetings, in terms of both time and authority. In Nortown, the LEA is particularly dominant; in Tolside, it is the EAZ officials who are mostly to the fore. The specialist knowledge, expertise, and bureaucratic power invested in these roles, in addition to the informal power the officials are able to wield behind the scenes, allows them literally to set the agenda, decide which issues to report back, and frame the nature of any discussion. To some extent, this may be due to external constraints, in particular the speed with which EAZs have been implemented. In addition, because EAZs are given priority access to numerous other new government initiatives, EAZ directors are under pressure not only to apply for any and all additional resources, but also to spend them within exceedingly tight time frames.

This not only reduces the scope for developing autonomous local initiatives, but also seems to encourage immediate and unilateral action by EAZ directors on a fairly regular basis. More broadly, it is clear that EAF partners who are not educationalists often find it difficult to contribute to what is essentially a very school-focused agenda. When Forum members were asked to describe the issues that generated the most discussion in EAF meetings, we were told that financial details such as arrangements for Value Added Taxation (VAT), mechanisms for tracking and reimbursing schools for EAZ initiatives, and deadlines for spending fundings had dominated the time available. These discussions were typically not seen to be of direct interest—or relevance—to non-school-based partners. External partners were also very aware of their own lack of educational expertise and were frequently unsure as to how they could play a more active role on their EAF:

> Eighty percent of the people there are heads [so] there are going to be an awful lot of issues that aren't really issues that are of any great concern to the bit of work we are doing, other than in very general, interest terms. So you always feel marginalized in that sense, because, particularly if you are not coming from an education background. (Voluntary sector partner, Tolside)

> I think it's also a weakness that the vast majority of the Forum, myself included, don't have that first-hand experience where we could, we would feel comfortable contributing to something we shouldn't contribute to, not our natural working area. (Business partner, Nortown)

The notion of partnership outlined in EAZ guidance emphasizes "shared ownership" of EAZ successes or failures, but says very little on the specific roles that partners are expected to play or how different types of knowledge and expertise might be utilized. In practice, this means that unless Forum business addresses core concerns of particular EAF members—or the initiatives in which they are directly involved—they seem unlikely to contribute actively or make time to attend regular meetings.

In saying this, we are not suggesting that power and influence remain constant either over time or even over the course of a given meeting. In Nortown, business partners contribute to discussions on a regular basis, usually in the form of raising issues and occasionally challenging the contributions of the LEA representatives and the zone director. Their voice, however, is not a privileged one. Even so, the relatively high level of involvement in EAF and executive board discussions by the business (Nortown) and headteacher (Tolside) partners far outweighs that made by the community representatives, including parents, whose contributions are minimal in both zones. This finding replicates what is known of the workings of school-governing bodies (Deem et al., 1995). As such, while the problem of involving "lay" people is not unique to EAZs, their Forums have yet to find ways of overcoming the barriers created by professional discourse. A Tolside parent representative recalled: "a fair amount of jargon is talked, there are many things in acronyms, whatever, and I didn't always know what that meant."

Despite the patterns we have identified in relation to differing levels of participation by the various EAZ partners, it is difficult to map the overall relative influence of these partners from our existing observational data. As the following quote from a Nortown parent governor illustrates, the apparent reliance on networking outside of EAF and executive board meetings privileges some EAZ partners over others: "people had their own agenda . . . Me being what they called 'lone parent representative' didn't have the connection between people." Parents and other community representatives who lack access to formal and informal networks are particularly disadvantaged in situations where strategic and operational decisions are made outside official EAZ meetings.

Our findings point to a lack of internal democracy within the two EAZs and lead to a concern that EAF meetings may serve more as an exercise in impression management than as a vehicle for frank and open debate of strategic direction. More generally, our research highlights the way in which the rhetoric of "partnership" serves to downplay the differing interests of various partners and mask tensions that include inequalities of power and resources (Atkinson, 1999; Hastings, 1999; Taylor, 2002).

Public–Private Partnerships: The Nature and Impact of Private Sector Investment and Involvement in Schools

EAZ policy texts may give the impression that EAZs represent a significant development in the history of private sector involvement in education, but in practice the vast majority of established zones have been LEA-initiated and DfEE claims about the numbers of "business-led" zones have been overstated (NUT, 2000; Hallgarten and Watling, 2000, 2001). Although "for-profit" education businesses, such as Nord Anglia and Edison, did express an early interest in the management of zones, this interest quickly evaporated when it became apparent that the government did not intend companies to make a profit from their involvement as EAZ partners. In addition, although a variety of private sector partners have been drawn into EAZs—ranging from multinational management consultancies, supermarkets, and professional football clubs to small local businesses—it is perhaps significant that the only private company involved in leading a first-round zone is a not-for-profit education company, and that two out of the three "business-led" second-round zones are led by trusts set up by companies for educational purposes. Moreover, so far none of the governing bodies of schools in either the first- or second-round EAZs have ceded their powers to their local EAF and nationally it is clear that "the amount of genuinely new money that the private sector has contributed to EAZs has undershot expectations" (Hallgarten and Watling, 2000, p. 26). Three years after the launch of the policy, only 12 out of 73 EAZs had managed to secure the expected £250,000 per year from the private sector and most of this sponsorship has taken the form of "in kind" contributions rather than cash (Mansell, 2001).

Here we present data drawn from three first-round EAZs—Brickly, Seaham, and Wellford—in order to explore what private sector investment and involvement in these EAZs has meant in practice (see table 4.2). The EAZs were selected for study because they claimed to be piloting particularly interesting or "radical" curriculum innovations, some in conjunction with private sector partners. Our analysis is based mostly on interviews with headteachers and teachers, conversations with pupils, and classroom observations conducted in four schools within each EAZ. Each zone had been operating for around a year prior to our fieldwork, which involved spending up to a week in each of the 12 schools visited. This fieldwork is supplemented by interviews with a sample of 24 zone "partners" across the three EAZs. At the time the research was conducted, EAZ-related business contributions took three principal forms within the 12 case study schools.

There are large discrepancies in the levels of private sector sponsorship obtained by the three EAZs. This is indicative of a national context in which access to private sector funding favors zones with the good fortune to be located near large, sponsorship-minded local businesses rather than necessarily the areas with the greatest need. In fact, the fundraising efforts of zones in the most economically distressed areas are likely to be hampered by the dearth of large local employers. However, the presence of large national or multinational EAZ partners in itself is no guarantee that their contribution will be proportionally large. Large-scale business sponsorship by one company seems to act as a spur to others; conversely, a failure to attract big contributions appears to be self-perpetuating. An important consequence of this local variation, therefore, is a massive disparity between the amounts of private sponsorship that different zones have been able to raise. Amongst the 25 first-round zones, in their first financial year, one raised £400,000 in private sector cash or "kind" contributions while nine raised less than £20,000 (NUT, 2000). Three years after the policy was first introduced, significant variations in private sector sponsorship levels were found across both first- and second-round EAZs (Mansell, 2001).

Private sector contributions of all kinds are also subject to fluctuation over time because of the vagaries of local economies. In at least two of our case study zones the prevailing local economic climate reduced the capacity of the various partners to contribute even over the first year. "Community" budgets tend to be first in line for reductions at times of hardship. Changes in company personnel can also lead to fluctuations in private sector contributions. The voluntary involvement of partner organizations usually depends on the commitment of individual

Table 4.2 Private sector contributions in EAZ schools

Type of contribution	Details
Cash	Private sector cash contributions make up a very small proportion of overall business sponsorship in each of the three EAZs. In the first financial year, Wellford obtained a negligible sum from direct cash donations, Seaham gained around £10,000 worth of contributions from six different partners and Brickly secured a substantial sum from one private sector partner.
Equipment and materials	Typical contributions include the provision of Information and Communication Technology (ICT) equipment and sponsored curriculum materials to support curriculum enrichment schemes. The most significant donation of this type is in Brickly, where a partnership with an ICT company allowed the zone to provide all schools with new ICT equipment. On a much smaller scale, partnerships with a football club and a supermarket in Wellford led to the provision of (football-related) activity books and (supermarket-related) worksheets for some zone pupils.
Human resources	A small number of schemes have engaged private sector companies in the provision of some sort of service to zone schools. Typically, such initiatives involve placements of private sector personnel in schools and/or pupil visits to non- school premises. Several of these initiatives have culminated in "events" that are frequently either one off or short term and small scale, involving only a small proportion of zone pupils.
Curriculum enrichment	Partnerships with professional football clubs, local artists, and drama groups have led to "football fun days," arts-related activities and visits to and from theater groups in all three zones. All zones have also provided a small number of additional work experience placements for zone pupils in some or all of their secondary schools, alongside the introduction of vocational courses (GNVQs) or an increase in the number of GNVQs offered.
Managerial services	A management consultancy has provided managerial and financial advice to the zone director in Seaham. In Wellford, a local company supplied discounted support for zone schools implementing a new system of management/professional review, while Brickly contracted out the task of monitoring pupil absences to a private company in a number of zone schools.
Mentoring	A small number of private sector volunteers have been involved in listening to children read in some local primary schools in Seaham and Wellford and, at the time of our initial fieldwork, all three zones were intending to establish a business mentoring scheme involving some zone headteachers and/or pupils.

members of staff. When these individuals change jobs or, as often seems to happen, they get internal promotions, the links weaken and may even fade altogether.

While levels of business contributions in two of our three study zones were amongst the highest nationally, these zones were not immune from the types of creative accounting practices that Hallgarten and Watling (2000, 2001) identified as being widespread. These include the overvaluation of "in-kind" contributions and the recycling of public, private, and voluntary sector money by the relabeling of existing schemes as "EAZ-related." Indeed the Seaham director admitted to meeting his sponsorship targets for the first year in large part by relabeling a host of preexisting initiatives.

New partnerships take some time to establish—and perhaps even greater effort to maintain—as the high turnover of EAF membership and the short-term nature of many zone initiatives attests. So far, as we discuss here, EAZ-related public–private partnerships have had an apparently limited direct effect on the overt curriculum of the 12 zone schools we studied. But this is not to say that they are having no impact on the wider teaching and learning environment. It is clear that some companies are using EAZs to pilot products and services that they aim to introduce elsewhere. In Brickly, for example, a company that was contracted to provide an attendance monitoring service has since expanded the service to cover non-EAZ schools in this district, while another company that donated ICT equipment to the zone's schools promptly received large orders from other EAZs and, more recently, from non-EAZ schools. Similarly, a management consultancy in Wellford briefly[5] doubled the number of local schools it was working in by securing an EAZ contract, which it was then able to use as a "selling point" elsewhere.

However, given the broader context of widespread business involvement in English state schooling, it is difficult to claim that these "new" public–private partnerships are having a major additional effect on the curriculum of any of the 12 zone schools. Indeed, few of the initiatives identified within the three EAZs involve "partnership" at all—if this implies more than straightforward commercial sponsorship within a public sector organization. Moreover, almost all of the schemes that did go beyond straightforward sponsorship were part of already established national or local schemes. Whilst some schools took part in new initiatives that placed business personnel or commercially sponsored curriculum materials in schools, the scale of direct business involvement on curriculum initiatives, attendance at meetings, visits to schools, mentoring or work placements was relatively small. The following four vignettes provide an idea of the nature and scale of EAZ-related business involvement in schools in one of our case study EAZs—Wellford.

Example One   Fast-Food Reading Mentors

This initiative, which places "Fast Food Reading Mentors" into zone schools, is part of a wider national scheme run by a large multinational. Only one of the three primary schools we visited in this zone was involved in the initiative and the scale is small, affecting only eight children in year five. Two volunteer reading mentors visit the school once a week for an hour to listen to four children read for 30 minutes each. The headteacher selected "competent readers" who she felt "would benefit from additional encouragement" to take part in the scheme; she was adamant that untrained volunteers would not be suitable for working with "less able" children.

In the lesson we observed it was clear that the children involved in this scheme enjoyed the special attention and time away from their regular class. At the same time, it is possible to question whether the volunteers had appropriate skills for this type of work. For example, one mentor periodically interrupted one of the girls she worked with to "correct" her (working class, regional) pronunciation: "NOT bu'er, BUTTER..." The other mentor was not himself a very confident reader and he stumbled repeatedly in his efforts to read aloud to the children he worked with.

Example Two   Reading and Football

The Reading and Football initiative is part of a wider national scheme involving professional football clubs around the country. We were able to observe a half-day "Reading and Football" session in one Wellford-year 5–6 class. The session we observed began with three footballers handing out "Football Funbooks" to all the pupils. The activity books, which are sponsored by a biscuit manufacturer, have an undemanding breezy style, with lots of cartoons and pictures of the sponsors' logo on every page. Activities include crosswords and football quizzes as well as short writing and mathematics tasks, for example, "Can you work out how many [brand name biscuits] you would need to build a pyramid with a base of 10?"

The children began by filling in details such as their name, nearest football club, and team they support. The three footballers then asked volunteers to read out some of the questions, the answers to which the pupils all filled in at the same time. Both the questions and illustrations are highly gendered; all but one question related to male players and male teams. The accompanying imagery shows girls in a variety of passive poses: girl sits looking bored on sofa facing TV while boy jumps up and down shouting; boy kicks ball while girl holds sign directing him to the goal.

> *Example Three* Supermarket Maths Trail
>
> The "maths trail" was a Wellford-wide initiative that involved a day visit to a local supermarket (part of a national chain). At Kite Hill Primary, 50 percent of their reception year pupils participated in the day trip. Teachers selected the "higher ability" pupils (aged four) whom they "thought would benefit" to participate in the outing. The school was given supply cover for the accompanying teachers, transport costs, and snacks for the children at the supermarket café. Once at the supermarket, children were given a glossy worksheet to complete as they wandered around the store in groups.
>
> The worksheet asks questions involving weights, numbers, and money, accompanied by pictures of "Supermarket" brand products. In the final section, pictures of fruit, chocolate bars and the like, are accompanied by "price tags" that children are instructed to add together. While there is a disclaimer about the prices displayed at the bottom of the page, all the price tags are severely undervalued.

> *Example Four*: Retail Work Experience
>
> This is the only example of an EAZ-related work experience placement at Millbank Community School. As part of the retail multinational "Family Stores" contribution to Wellford EAZ, it has provided a ten-day work experience placement for six local year-ten pupils. This school selected two "sensible and trustworthy" girls who had volunteered to participate primarily in order to enhance their future opportunities of gaining part-time work when they go on to pursue further study. Awareness that pupils were "representing the school" apparently led teachers to discard the notion of choosing disaffected or challenging pupils.

These examples are fairly illustrative of the kinds of "curriculum enrichment" activities in which the business sponsors of our case study zones are involved. There are several key points to note here. First, what is perhaps most striking is the banality of business involvement. The businesses involved in our case study EAZs have certainly not so far demonstrated the energy, creativity, and know-how to transform radically the provision of education in socially disadvantaged areas that at least one government official has claimed on their behalf (Barber, 1998a, cited in Jones and Bird, 2000). Second, whilst the far-reaching commercialization of the school curriculum feared by critics of the policy has not emerged, we have identified some isolated examples of the use of curriculum materials that are clearly aimed at promoting the products of their commercial sponsor. Third, it appears that through their eagerness to attract additional resources to the school, teachers may lose a degree of control over the types of curriculum materials that their children are exposed to. Both in the case of Wellford's "Football and Reading" initiative, which used highly gendered imagery, and its "Supermarket Maths Trail," with its promotional orientation, the class teacher did not see the material before it was used with her students. As the "Fast-Food Reading" example illustrates, we also observed the work of business reading mentors whose competence to assist in the learning process was questionable. Second-round zones that are eager to attract DfES (Department for Education and Skills) "matched" funding to support core services, are in an even poorer position than the first-round zones we studied to negotiate for high quality, relevant private sector contributions and instead are likely to feel under pressure to accept whatever types of support local businesses are willing to offer. Finally, due to the relatively small amounts of money allocated to any given EAZ innovation, almost all zone initiatives whether school-based or zone-wide involve some form of targeting or selection. As the examples we have given show, where these initiatives have direct private sector involvement, they tend to target relatively "able" or well-motivated pupils. This is a finding that raises questions about whether the interventions are reaching those most in need, and appears to reflect both the desire of businesses to associate themselves with "high achievers" and "headteachers" consciousness of how best to represent their school to the outside community.

Multiagency and Interschool Partnerships: Barriers to Joined-Up Working

In addition to community-based partnerships and partnerships between the public and private sectors, EAZs were meant to promote partnerships between different welfare sectors in order to reduce the fragmentation of welfare provision and to facilitate interschool partnerships. However, although interviews with zone partners have indicated that new multiagency initiatives are more numerous and significant than our early fieldwork has suggested, we encountered only a few isolated and small-scale cross-welfare initiatives in the 12 schools we studied in depth. The most significant of these initiatives has placed a part-time counselor from a mental health trust in one of the four case study Seaham schools. Other initiatives that have been credited as "EAZ-related" include the joint Health Action Zone/EAZ funding of breakfast clubs in Wellford, and the involvement of a large voluntary agency in the management and provision of a childcare project in one Wellford school (this project is not EAZ funded and predates the zone). We have not come across any multiagency initiatives of any type in our four Brickly case study schools. So how might we explain the apparent paucity of multiagency initiatives? As several commentators have noted, interagency boundaries have historically proved to be remarkably resilient (Wilkinson and Appelbee, 1999; Riddell and Tett, 2001; Webb and Vulliamy, 2001; Glendinning et al., 2002). But there are also, paradoxically, significant barriers to multiagency working and inter-school partnerships emanating from elements of the EAZ policy itself as well as the wider policy context in which EAZs are located. Here we will focus on the ways in which the bidding culture and the quasi-market within which EAZs are enmeshed seem to be inhibiting joined-up working.

Zone schools are frequently in receipt of funds from a range of sources in addition to the resources they get by virtue of their EAZ status. EAZ schools are given "priority access" to funding from a number of other DfES initiatives and some also receive funds from the Single Regeneration Budget, the New Opportunities Fund, and the European Social Fund. Having constantly to bid for additional resources places a significant time burden on both EAZ directors and headteachers, and it leads to inequalities in access to resources within zones, as levels of school-based funding are at least partially dependent on headteacher expertise, time, and inclination to bid for zone resources available from a menu of projects on offer. As one Wellford headteacher explained:

> Literally we sort of got what we bid for, and that was perhaps an inequality because some people hadn't come from a bidding culture and they didn't really have a clue what to bid for, whereas others of us had a list as long as your arm, you know [laughs]. I mean, some bids I didn't even have to re-write, they were bids that I'd already put out for other money . . . all our initial projects were bids. (Headteacher, Wellford)

Recognition of the importance of additional fundraising has led at least two of the five case study zones discussed in this chapter (Tolside and Brickly) to hire an "income generator" to specialize in attracting funding from public, voluntary, and private sector sources. Similarly, a secondary school in Seaham has appointed a partially EAZ-funded deputy headteacher to generate additional income and manage externally funded projects.

Our early evidence suggests that, in practice, the bidding culture within which zones are enmeshed may actually increase the fragmentation of provision, because the coexistence of many small-scale, tightly bounded projects is difficult to coordinate and has in some instances led to a confusion of responsibilities between EAZ, LEA, and school personnel. The involvement of several public, voluntary, and private sector organizations working on different projects with very similar objectives may further exacerbate this situation, as we found limited evidence of meaningful collaboration between service providers.

There are additional tensions within the national policy agenda that are likely to hamper both cooperation between schools and the development of "joined-up working" more generally. As Clarence and Painter (1998, p. 15) argue: "[t]he collaborative discourse of New Labour is countered by another, conflicting and contradictory discourse of centralization and demands for performance." New Labour has, for example, extended some of the managerialist techniques associated with previous Conservative administrations. In English schools, children aged seven, eleven and fourteen are required to sit for national tests known at Standardised Attainment Tasks (SATs) in mathematics, English, and science. At each level there are national performance thresholds that children are expected to meet, and both LEAs and individual schools are required to set targets against which their progress can be measured. EAZs, in turn, are obliged to set even more stringent targets.

Since judgments about the success or failure of individual schools, LEAs, and EAZs are based, in large part, on the their ability to meet these performance targets, this may help to explain why concrete multi-sector initiatives have, in practice, been so peripheral to core EAZ business. Any effort to build local partnerships is likely to be time-consuming—and as many headteachers have pointed out, the EAZ timescale is relatively short—while the "pressure to perform" is high:

> I think there are two philosophies really, I think that there's the philosophy of—this is my school and I'm going to get what I can for my school, and I'm going to think of my school. And there are other people who agree with the philosophy that we share amongst other people. And it is a very difficult one for heads, isn't it? Because at the end of the day you are responsible for your school. I am not responsible for the school over the road. I'm just responsible for my school. So it's kind of like getting away from that insular way and it's looking out isn't it really? (Headteacher, Nortown)

> I suppose all headteachers, when it comes down to the money will say— oh, we want, you know, I want what's best for my school . . . but as far as the EAZ is concerned, they appear to be less interested in discussing the plans for the good of the whole of the EAZ than for what they can get out of it for their particular [school cluster]. (Headteacher, Tolside)

Within a quasi-market context, it seems that the benefits of collaboration are unclear, while the risks may be perceived as considerable. Accordingly, although a number of zone headteachers have spoken positively of the "cooperative climate" fostered within their EAZ, this appears to have had very little impact on classroom practices. In addition, the majority of EAZ learning support has been used to boost the ability of schools to meet their own individual performance targets. The most resource-effective way to do this of course is not to target the most disadvantaged pupils, but to focus on those who are deemed to be capable of meeting an assessment benchmark (Gewirtz et al., 1995; Gillborn and Youdell, 2000). In Wellford, for example, one of the primary schools we studied held an "invitation only" Easter school for "borderline pupils" immediately prior to the compulsory national tests (SATs). Similarly in Brickly, one of the primary schools was selected for mentoring support pupils who were "just below" meeting the level four SATs performance threshold, while the secondary school allocated resources to allow senior members of staff to work with "able but underachieving pupils" who were "just below" gaining the threshold number of five A-C grades in the General Certificate of Secondary Education (GCSE).[6] There is evidence that targeting of resources on students deemed to be on the borderline of assessment benchmarks is a widespread practice (Gillborn and Youdell, 2000). However, there is a particularly strong incentive for EAZ schools, keen to secure funding beyond the initial three-year period, to focus on those students whose assessment results are most likely to affect their ability to reach school-based performance targets. While such strategies may be an "efficient" use of EAZ resources for schools in competition for local pupils, it does raise significant equity concerns.

Conclusions

The English EAZs were introduced as new kinds of partnership that would bring together groups of schools, parents, and local private, voluntary, and community organizations, including agencies from different welfare sectors, to tackle disadvantage and raise standards in schools (DfEE, 1997). Our evidence suggests that in assessing the nature and operation of contemporary partnerships it is important to distinguish between what is stated in policy texts and zone practices. As Janet Newman (2002, p. 81) argues, "[a]lthough the discourse of partnership signifies equality of power, shared values and the establishment of common agendas and goals, the organizational reality tends to be very different." Our research shows that the notion of EAZs as "partnership organizations" often tends to obfuscate more than it illuminates.

The success of any multi-sector partnership is in part reliant on its ability to involve diverse agencies. However, the distribution of partners within EAZs is more uneven, and active involvement more difficult to secure, than the rhetoric of partnership suggests. Certainly at the strategic level there is limited indication of real, "shared ownership" of EAZ decision-making, and rather greater evidence that EAFs have reinforced existing local power relations. As presently constituted, the EAFs we have been observing do not appear to provide "a very progressive mode of civic association" as envisaged by some early commentators (Halpin, 1999, p. 234). Nor currently are there signs that the EAZs in question are beginning to provide "the capability citizens need for the task of regenerating civil society" (Ranson, 2000, p. 263). This issue has particular significance, given the important symbolic role EAFs were meant to play in legitimizing EAZs as new, more inclusive local partnerships for planning and delivering local services. If anything, what we discern is a further consolidation of managerial interests, notably those represented by LEAs, EAZ managers, and the headteachers of participating schools.

Our research has also drawn attention to the relatively small-scale, limited, and uneven nature of private sector investment and involvement in zones. Uneven access to private sector (and other EAZ resources) occurs on a number of levels—between zones, between schools in the same zone, and even *inside* schools. In particular, schemes involving the private sector have tended to target a small proportion of students, often those deemed to be the most able or motivated. The issue of unevenness does not relate just to geographical or institutional distribution, it also relates to sustainability over time with fluctuations in private sector contributions arising from changing economic fortunes of the companies involved and changes in personnel. These problems raise issues as to whether the already low levels of business involvement will be sustained in the near future—let alone beyond the lifetime of the zones. Perhaps more importantly, the nature and limited scale of private sector initiatives casts doubt on the suggestion that business involvement of the kind in evidence in the zones we have studied has the capacity to revitalize education in socially disadvantaged areas (see Barber, 1998a, cited in Jones and Bird, 2000). It is perhaps as a consequence of this that the far-reaching commercialization of the school curriculum feared by critics of the policy (STA, 1998; Wilby, 1998) has not materialized. Nor have we uncovered evidence to suggest that business involvement in EAZ activity is leading to a diminution in the role of other interest groups.

As regards multi-sector initiatives, we have little evidence of these having a concrete impact at the level of individual zone schools at the time of our fieldwork. In part, this appears to be a function of the targets against which the success of each EAZ is measured. Despite DfEE guidance that indicated that EAZ partnerships would be free to develop their own priorities, plans, and initiatives to suit local circumstances, in practice the level of resources devolved to individual EAZs is relatively small (Plewis, 1998), and the government has tightly defined the expected policy outputs. This does not simply influence the aims to which zones are expected to work, but helps to determine the strategies that they feel able to pursue. We suggest that a focus on easily measurable, short-term targets has discouraged the implementation of longer-term multi-sector and community-based initiatives in favor of educational interventions that may more directly act to raise pupil performance in school-based tests. Difficulties in establishing multi-sector working have further been exacerbated by the bidding culture associated with EAZs, which militates against coordinated action at both strategic and operational levels. The bidding culture can both discourage consultation between zone partners and promote a fragmentation of provision, leading to an unwieldy proliferation of small-scale, short-term projects that are difficult to manage and that may lead to a confusion of responsibilities between the different agencies involved.

However it is not only multi-sector working that is difficult to achieve within a competitive context. Despite EAZ efforts to encourage collaboration between schools, pressure to meet performance targets and competition for pupils within local education markets may ultimately hamper interschool cooperation. Market forces also encourage schools to target resources on those students deemed to be at the borderline of assessment benchmarks, particularly schools that are subject to additional EAZ performance targets.

The problems of fragmentation, confusion, and duplication associated with area-based initiatives—and partnership working in general—have recently been recognized in a variety of U.K. government commissioned reports (DETR, 2000; GIDA, 2000; PIU, 2000; SEU, 2001; NRU, 2002). In response, the Labour Government has recently signaled a shift in emphasis away from "special" zonal initiatives and toward the mainstreaming of programs geared toward reducing the gaps between the most deprived neighborhoods and the rest of England.

Following the publication of two reports indicating that EAZs have had mixed success in meeting their policy objectives (CELSI, 2000; Ofsted, 2000), the government announced that no further large EAZs are to be introduced. However, other new forms of public–private partnership have emerged since the launch of the EAZ policy and these are likely to expand in the future. These include the privatization of national government services, such as the administration of performance-related pay; the outsourcing of LEA services to private companies; the involvement of private firms (some of whom have minimal experience in education) in running "failing" LEAs; and fixed-term contracts with private (for-profit and not-for-profit) companies and charities to run state schools (Hatcher, 2001). At the same time, over a hundred small EAZs have been established as part of the Excellence in Cities[7] initiative. Like the large EAZs discussed in this chapter, small EAZs are meant to have a multi-sector focus and to secure private sector sponsorship, although unlike the large EAZs they operate within traditional LEA structures. It is yet to be seen whether those who are promoting and implementing these new forms of "partnership" have learnt any of the lessons of the large EAZ experiment—or indeed of previous attempts to coordinate action across welfare sectors or harness private sector funds and energies to the provision of state schooling. Critics of zones may argue that uneven coverage, lack of sustainability, weak accountability, and undemocratic decision-making processes are endemic features of cross-sector partnerships. The challenge facing the United Kingdom and other governments committed to promoting partnership models of governance is to demonstrate that they are not—and that such partnerships can indeed provide a viable framework for the effective, democratic, and equitable delivery of welfare services.

Notes

1. This is a revised and extended version of a chapter published in Glendinning, Powell, and Rummery (2002). We are grateful to Caroline Glendinning, Martin Powell, Kirstein Rummery, and the Policy Press for permission to reuse material from that chapter here.
2. Other area based initiatives introduced by New Labour in their first term included Health Action Zones, Employment Zones, Sure Start, and the New Deal for Communities.
3. ESRC reference: R000238046.
4. The DfEE was renamed the Department for Education and Skills (DfES) following the general election in June 2001. In this chapter we refer to the DfEE when discussing the period up to June 2001, and DfES when referring to events from June 2001 onward.
5. In fact, this "partnership" dissolved in less than a year when EAF members questioned why the contract had not been sent out for tender. The company involved (who had been offering the services to the zone at what they considered to be a significant discount) had their EAZ funding withdrawn, and subsequently resigned as EAZ partners—vowing never to become involved in similar initiatives in the future.
6. The General Certificate of Secondary Education (GCSE) is the principal means of assessing pupil attainment at the end of compulsory secondary education in England.

7. Excellence in Cities (EiC) was launched in September 1999 as a major new initiative to address social disadvantage across urban areas of England. Although the policy can be seen in some ways as a successor to the EAZ initiative, EiC money has been given to urban LEAs and then channelled directly into schools. The policy has several key strands that link closely to mainstream policy initiatives. These include the provision of high-tech city learning centers and learning support centers, expansion of the specialist and beacon school programs, development of gifted and talented schemes, the introduction of learning mentors, and the establishment of mini EAZs.

References

6, Perri. (1997). *Holistic Government*. London: Demos.
Alexander, A. (November/December 1991). "Managing Fragmentation—Democracy, Accountability and the Future of Local Government." *Local Government Studies*, 63–76.
Atkinson, R. (1999). "Discourses of Partnership and Empowerment in Contemporary British Urban Regeneration." *Urban Studies*, 36(1), 59–72.
Barber, M. (January 7, 1998a). *The Guardian*.
Barber, M. (June 23, 1998b). "The Four Pillars of a First-Rate Education." *The Independent*.
Blair, T. (1998). *The Third Way: New Politics for the New Century*. London: Fabian Society.
Byers, S. (June 26, 1998). *The Guardian*.
Cabinet Office (1999). *Modernising Government*. London: the Stationery Office, from, http://www.cabinet-office.gov.uk/moderngov/1999/whitepaper/cover.htm.
CELSI (2000). *Evidence of Progress: An Independent Review of Evaluation Activities in Education Action Zones*. Canterbury: Canterbury Christ Church University College.
Clarence, E. and Painter, C. (1998). "Public Services under New Labour: Collaborative Discourses and Local Networking." *Public Policy and Administration*, 13(3), 8–22.
Deem, R., Brehoney, K., and Heath, S. (1995). *Active Citizenship and the Governing of Schools*. Buckingham: Open University Press.
DETR (Department of the Environment, Transport and Regions) (2000). *Joining it up Locally*. Report of Policy Action Team 17, London: DETR, from, http://www.local-regions.odpm.gov.uk/pat17/jiul17.pdf.
DfEE (Department for Education and Employment) (1997). *Education Action Zones: An Introduction*. London: DfEE.
DfEE (1998a). *Guidance on Completing the Form*. London: DfEE.
DfEE (1999). *Meet the Challenge*. London: DfEE.
DH (Department of Health) (1998). *Partnership in Action: New Opportunities for Joint Working Between Health and Social Services*. London: DH, from, http://www.doh.gov.uk/pia.htm.

Dickson, M., Gewirtz, S., Halpin, D., Power, S., and Whitty, G. (2001). "Education Action Zones and Democratic Participation." *School Leadership and Management* 21 (2), 169–182.
Gewirtz, S. (1999). "Education Action Zones: Emblems of the Third Way." In H. Dean and R. Woods (Eds.), *Social Policy Review* 11 (pp.145–165). Bristol: Social Policy Association/Policy Press.
Gewirtz, S., Ball, S.J., and Bowe, R. (1995). *Markets, Choice and Equity in Education*. Buckingham: Open University Press.
(GIDA) (2000). *Government Interventions in Deprived Areas: Report of the SR 2000c Cross Cutting Review*, from, http://www.hm-treasury.gov.uk/mediastore/otherfiles/518.pdf.
Gillborn, D. and Youdell, D. (2000). *Rationing Education: Policy, Practice, Reform and Equity*. Buckingham: Open University Press.
Glendinning, C., Powell, M., and Rummery, K. (Eds.) (2002). *Partnerships, New Labour and the Governance of Welfare*. Bristol: Policy Press.
Hallgarten, J. and Watling, R. (2000). "Zones of Contention." In R. Lissauer and P. Robinson (Eds.), *A Learning Process: Public Private Partnerships in Education* (pp. 22–43). London: IPPR.
Hallgarten, J. and Watling, R. (2001). "Buying Power: The Role of the Private Sector in Education Action Zones." *School Leadership and Management*, 21(2), 143–158.
Hastings, A. (1999). "Analysing Power Relations in Partnerships: Is there a Role for Discourse Analysis?" *Urban Studies*, 36(1), 91–106.
Halpin, D. (1999). "Democracy, Inclusive Schooling and the Politics of Education." *International Journal of Inclusive Education*, 3(3), 225–238.
Hatcher, R. (1998). "Profiting from Schools: Business and Education Action Zones." *Education & Social Justice*, 1(1), 9–16.
Hatcher, R. (2001). "Getting Down to Business: Schooling in the Globalised Economy." *Education & Social Justice*, 3(2), 45–59.
Home Office (1998). *Compact on Relations Between Government and the Voluntary and Community Sector in England: Getting it Right Together*. London: Home Office, from http://www.homeoffice.gov.uk/vcu/compact.pdf.
Jones, K. and Bird, K. (2000). "Partnership as Strategy: Public–Private Relations in Education Action Zones." *British Educational Research Journal*, 26(4), 491–506.
Mansell, W. (July 27, 2001). "Private Sector Cold-Shoulders Fledgling EAZs." *Times Educational Supplement*, 2.
NUT (National Union of Teachers) (2000). *An Analysis of First Round EAZ Accounts 1998–1999*. London: NUT.
NRU (Neighbourhood Renewal Unit) (2002). *Collaboration and Co-ordination in Area-Based Initiatives*. London: DETR.
Newman, J. (2002). "The New Public Management, Modernization and Institutional Change: Disruptions, Disjunctures and Dilemmas." In K. McLaughlin., S.P. Osbourne, and E. Ferlie (Eds.), *New Public Management: Current Trends and Future Prospects* (pp. 77–91). London: Routledge.
Ofsted (Office for Standards in Education) (2001). *Education Action Zones: Commentary on the First Six Zone Inspections*. Ofsted.

PIU (Performance and Innovation Unit) (2000). *Wiring it up: Whitehall's Management of Cross-Cutting Policies and Services*. London: Cabinet Office, from, http://www.cabinet-office.gov.uk/innovation/2000/wiring/coiwire.pdf.

Plewis, I. (1998). "Inequalities, Targets and Zones." *New Economy*, 5(2), 104–108.

Powell, M. and Glendinning, C. (2002). "Introduction." In C. Glendinning., M. Powell., and K. Rummery (Eds.), *Partnerships, New Labour and the Governance of Welfare* (pp. 1–14). Bristol: Policy Press.

Power, S. (2001). " 'Joined Up Thinking?' Inter-Agency Partnerships in Education Action Zones." In S. Riddell and L. Tett (Eds.), *Education, Social Justice and Inter-Agency Working: Joined Up or Fractured Policy?* (pp. 14–28). London: Routledge.

Power, S. and Gewirtz, S. (2001). "Reading Education Action Zones," *Journal of Education Policy*, 16(1), 39–51.

Power, S. and Whitty, G. (1999). "New Labor's Education Policy: First, Second or Third Way?" *Journal of Education Policy*, 14(5), 535–546.

Rafferty, F. (February 6, 1998). "Action Zones will Pilot New Ideas." *Times Educational Supplement*, 6.

Ranson, S. (2000). "Recognizing the Pedagogy of Voice in a Learning Community." *Educational Management and Administration*, 28(3), 263–279.

Riddell, S. and Tett, L. (Eds.) (2001). *Education, Social Justice and Inter-Agency Working: Joined Up or Fractured Policy?* London: Routledge.

SEU (Social Exclusion Unit) (2001). *A New Commitment to Neighbourhood Renewal: National Action Strategy Plan*. From, http://www.socialexclusionunit.gov.uk/publications/reports/html/action_plan/index.htm.

STA (Socialist Teachers Alliance) (1998). *Trojan Horses—Education Action Zones: The Case Against the Privatisation of Education*. London, STA.

Taylor, M. (2002). "The New Public Management and Social exclusion." In K. McLaughlin., S.P. Osbourne, and E. Ferlie (Eds.), *New Public Management: Current Trends and Future Prospects* (pp. 109–128). London: Routledge.

Tooley, J. (1998). "Education Action Zones." *Economic Affairs*, IEA.

Webb, R. and Vulliamy, G. (2001). "Joining up the Solutions: The Rhetoric and Practice of Inter-Agency Collaboration." *Children and Society*, 15, 315–332.

Wilby, P. (March 20, 1998). "This May be the End of the LEA Show." *New Statesman*, 20, 24–25.

Wilkinson, D. and Appelbee, E. (1999). *Implementing Holistic Government: Joined-Up Action on the Ground*. Bristol: The Policy Press.

CHAPTER 5
PARTNERSHIPS: THE COMMUNITY CONTEXT IN MIAMI

M. Yvette Baber and Kathryn M. Borman with Jennifer Avery and Edgar Amador[1]

The National Science Foundation (NSF) has developed a model of systemic change in mathematics and science that has been applied to urban schools around the United States in the last decade. In conceptualizing this model, NSF argued that both school and community resources are essential in securing long-term success in increasing student achievement (National Science Foundation, 2001). Although NSF reforms are principally aimed at developing constructivist approaches to teaching and learning in mathematics and science classrooms through professional development opportunities, the NSF model for systemic change posits multiple policy levers (Drivers)[2] for creating enduring reforms, particularly in classrooms located in the most economically challenged communities.

Stakeholders, partners, and other resources are critical elements in supporting sustainable and effective reform, and they are also essential in encouraging and supporting individual student achievement. The effectiveness of multiple policy inputs (Drivers) in stimulating educational reform was evaluated by researchers at the David C. Anchin Center at the University of South Florida during a three-year period from 1999 to 2001. Our evaluation included an investigation of how each of four sites had identified and created relationships with parents, community agencies, museums, businesses, and others to provide critical resources to augment and amplify effective classroom instruction. The objective here was to understand how stakeholder involvement affected the implementation of reform at the school, community, and district levels. A series of ethnographic studies in the four Urban

Table 5.1 Impact of USI on partner involvement

Partner descriptions of impact	Chicago		El Paso		Memphis		Miami		Total	
	N	%	N	%	N	%	N	%	N	%
Little to no impact	12	21	5	9	11	19	12	21	40	70
Involved in USI program/ received USI funding	1	2	1	2	9	16	6	10	17	30
Total number	13	23	6	11	20	35	18	31	57	100

Systemic Initiative (USI) sites (Chicago, El Paso, Memphis, and Miami) took place during the second year of the three-year project. During our community study research, these high schools and their surrounding communities were the focus of intensive data collection to investigate ways in which community-level resources and residents perceived their neighborhood-based assets and how these assets contributed to supporting students' development in and outside schools. The studies also chronicled neighborhood histories, identified the organizational and human assets and resources available to high school personnel to support educational reform, and generated descriptions of community-level resources already in place.

We considered stakeholders or partners as *anyone* or any organization that worked with a school to support students and educational achievement. The literature has, for years, worked to identify types and categories of parent involvement, separating parents' activities from partnerships or other formal arrangements developed by school districts (or schools) with businesses, larger agencies, or other entities than families.

According to 70 percent (40 of 57) of the community and parent stakeholders interviewed in the four site research project, the USI reforms in mathematics and science had little to no impact on their involvement with students or schools. This finding was both disturbing and startling although, given the division between parent involvement and other partners at most schools, not surprising. As shown in table 5.1, involvement was defined for the purpose of the study as whether or not the partner participated in a USI program or received funding from the USI to support programmatic activity with the schools. This definition was, admittedly, a narrow one, but it followed criteria used to measure "involvement" activities in earlier national and state reform efforts such as Goals 2000 (U.S. Department of Education, 1998).

Conceptual Framework

The conceptual framework for this study combined a model of neighborhood and community influences with an integrative model of "development in context" for understanding the ways in which neighborhoods and partnerships influence the education of urban youth. This model drew on the work of Aber, Gephart, Brooks-Gunn, and Connell (1997) that considered the effects of three types of contextual processes on individual outcomes: (1) neighborhood and community processes; (2) social and interpersonal processes; and (3) individual processes. These three processes are influenced by exogenous (social, political, economic) forces from the broader society, creating an environment that shapes and inexorably impacts students' developmental and educational outcomes (see figure 5.1). This model expands upon and clarifies the importance of NSF's Driver 4 (mobilization of stakeholders) by demonstrating how social and interpersonal processes, as well as individual actions, shape students' developmental and educational outcomes.

It is becoming increasingly important to include neighborhood and community processes in educational research to expand a dialogue about the critical role individual and organizational partners play in affecting student development. Prior research has focused on either societal or individual processes as these are linked to children's adaptation, growth, and educational achievement. When community processes have been studied, it is the school that is the focus rather than other institutions in the community. We must understand how community-based institutions and processes influence school outcomes if schools and school districts are to benefit from a full range of partnership involvement.

In addition, community influences historically have not been directly centered in institutions (such as schools) and have been omitted from many education reform frameworks (Aber et al., 1997). Early research was confined to empirical analysis of census tract or longitudinal survey data as in the case of studies including the Panel Study of Income Dynamics, the National Longitudinal Survey of Youth, and the National Survey of Families and Households. Recently, multilevel, multisite studies have been undertaken by the MacArthur Foundation Research Network on Successful Adolescent Development among Youth in High Risk Settings investigating contexts external to schools that influence youth development (Furstenberg and Hughes, 1995). These studies, including the latest in the series undertaken by Frank Furstenberg and his colleagues (1999), have concluded that, with respect to successful adolescent outcomes, variation in neighborhood

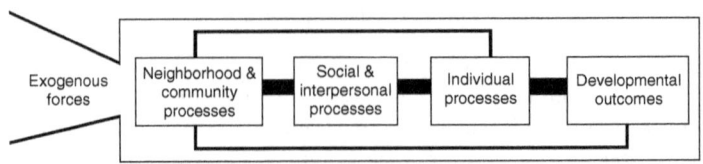

Figure 5.1 Model of development in context (Aber et al., 1997, p. 45)

influences is greater within rather than between neighborhoods. Far from being isolated in ghetto-like circumstances, most residents in a given urban neighborhood have the capacity to benefit from assets available city-wide, especially in large cities with good, inexpensive transportation systems. Nonetheless, in urban neighborhoods with greater assets, heaviest reliance is placed by families on institutional connections and social networks within their own neighborhoods, showing the importance of ties to community-based assets such as local public libraries in promoting adolescent success. The converse is true for families living in neighborhoods with the fewest resources. In these circumstances, the researchers found the heaviest dependence on family discipline strategies that emphasized restrictiveness and in-home investment (Furstenberg et al., 1999). These strategies have been found to lead to lower levels of successful adolescent outcomes. In this chapter, we will describe how these concepts played out in two communities in the Miami-Dade County Public Schools (MDCPS).

To fully mobilize all available individuals in the educational reform effort, attention must be given to the neighborhoods and communities in which schools are located and students reside. It is important to recognize that community characteristics vary. Subcultural codes and understandings subsequently also vary and these variations are transmitted to youth residing in neighborhoods everywhere. Unless neighborhood subcultures and their unique characteristics are recognized by educational policy makers, the out-of-school influences on student achievement and attainment will not be understood or given appropriate consideration in improving the odds that all students will benefit from educational reform programs. Important partners and school–community relationships will continue to be ignored or overlooked by district-level policy makers.

Researchers are also investigating the ways in which income and other social characteristics shape the involvement policies and practices of school districts as well as the practices of parents (Kessler-Sklar and Baker, 2000; Feuerstein, 2000; Lott, 2001). These studies will help to expand the definitions of partnerships and involvement as they provide information about the impact of social class and variations in cultural practices on the ways districts, communities, and parents interact with each other. We envision a period of redefinition, moving away from the middle-class perspectives and definitions of the past (and present) to include recognition of the diversity of activities that can be deemed partnerships or involvement with schools and students. This redefinition must be grounded in both qualitative and quantitative research to identify all facets of the diversity and all relationships that can be included in the description.

A discussion of the methodology applied to this qualitative segment of the USI evaluation research can be found at the end of this chapter.[3] A discussion of our study of two high schools and their relationships with community-level partners follows.

Miami-Dade County Public Schools

In this chapter, we highlight two senior high schools in the greater Miami area, describing the communities, ways in which individuals perceive and define their relationships with schools and the school district, and how these relationships constitute partnerships to support school reform and/or individual student achievement. By comparing and contrasting the two communities, we will discuss how larger societal processes (e.g., economic conditions, state laws, transnationalism, and migration) affect relationships between the school and the larger community and how these relationships affect opportunities for students. We selected the MDCPS because of the diversity prevalent throughout the district and because a variety of efforts have been initiated to support student achievement across the district. Our investigation uncovered a strong and positive relationship between after-school opportunities and achievement among Hispanic female students that speaks to the value of effective partner relationships at the local level. The two high schools are in the throes of important changes, but in vastly different ways. One has a population composed predominantly of first or second generation Haitian-descent students and struggles with the challenges of mobility, language, and the need to increase student achievement. The other is in the process of redefining itself as a diverse international school, serving students from many countries, while regaining its reputation for sound academics. After describing each of the two communities and schools, we will discuss the ways they have approached the concept of partnerships (individual and community relationships or involvement in district-wide initiatives to support student achievement) and the impact of larger societal processes on these relationships. We will look at how the resource utilization aligns with Furstenberg's (1999) description of resource use in urban neighborhoods. We will discuss the multilayered definitions of partnerships that evolved from our conversations with individuals in the schools, communities, and district. It is our hope to expand the concept of partnerships through this discussion and encourage educators and policy makers to consider forces outside of the school or home that shape and affect the presence and quality of those relationships.

Miami/Dade County: A Brief Description

The City of Miami is located in Dade County on the southeastern tip of the Florida peninsula. It boasts a semi-tropical climate, warm ocean waters, and sandy beaches, making it one of the major tourist areas in the United States, attracting almost 13 million vacationers a year (American Automobile Association, 2001). Miami-Dade County, with 2.1 million residents, is the largest "local" government in the southeastern United States. The county is governed by an Executive Mayor and the Miami-Dade Board of County Commissioners. According to the 2000 census, 362,470 people live in the city of Miami (U.S. Bureau of the Census, 2002). The Miami-Dade County Public School System is the fourth largest public school district in the nation, with over 370,000 students and 314 schools. This is a predominantly Hispanic community where over 238,351 residents (64 percent) identify themselves as Hispanic; 123,000 of that number are of Cuban descent. From 1990 to March 2000, 80,911 foreign-born persons entered the United States via Miami (U.S. Bureau of the Census, 2000). An increasing number of new immigrants are from South American countries including Argentina, Chile, and Brazil. (U.S. Bureau of the Census, retrieved December 29, 2002). Other cultural groups living in the Miami area include: African American, Anglo-American, Bahamian, Chinese, Cuban, Greek, Haitian, East Indian, Irish, Jewish, Miccosukee and other Native American groups, Nicaraguan, Puerto Rican, and Venezuelan.

One of the area's defining moments came in 1959 with Fidel Castro's takeover of Cuba, an event that led to a vast exodus of Cubans to the city of Miami. The business acumen of many exiles was a boon to the city and region's economy, while the vibrant Cuban culture brought new life to Miami. The United States sponsored "Freedom Flights," massive airlifts of Cubans to Miami from 1965 to 1973, delivered 150,000 Cubans to

America. By the 1980s, the large Cuban refugee population, whose county-wide numbers during the decade exceeded 600,000, was actively engaged in the political process, dominating the government of the City of Miami, as well as those of neighboring communities. Little Havana, the initial entry point for early waves of Cubans, had also become the destination for refugees from other Spanish-speaking countries in the hemisphere, especially Nicaragua. In Miami's northern sector, refugees from Haiti were pouring into Lemon City and transforming that bastion of old Miami into an Afro-West Indian community.

In the 1990s, Miami's place as a refugee haven put tremendous financial burdens upon the metropolitan area, making it one of the poorest in the United States. The refugees' competition with American-born blacks for entry-level jobs led to simmering tensions between immigrants and resentful residents of Liberty City, Brownsville, and other native black communities. Despite gains realized by Miami's African Americans in the aftermath of desegregation, poverty and crime remained disproportionately high in this group, and black anger over the perceived inequities and biases of the criminal justice system led to a series of searing riots in the 1980s and 1990s. Florida is a southern state, affected by the Jim Crow laws of the 1880s–1960s and a history of racial unrest. These racial tensions flared in 1989 when a Miami police officer shot and killed a young African American youth. Racial unrest ensued and intensified after he was found not guilty of homicide at his trial. Adding to Miami's problems at the end of the century were the city's notoriety as a haven for drugs, especially cocaine brought in from Latin America, the pervasive problem of crime, and relationships with Cuba. The Mariel Boatlift of 1980 transformed the demographic landscape of Miami, and especially Miami Beach, with the arrival of thousands of Cuban nationals. Now, immigrants (both legal and illegal) from many Caribbean and Latin American countries come in to Miami and Dade County. The area is seen as a receiving point for Spanish-speaking immigrants and those from the islands of the Caribbean due to its climate, multinational population, and proximity to homelands.

Education Reform
The MDCPS received generous support from the National Science Foundation (NSF) to implement a variety of educational reform programs across all grades and curricula. In the 1990s, the district received funding from the NSF for both the USI and the more recent mathematics and science reform program, the USP (Urban Systemic

Program). Both the USI and USP programs and practices incorporate the National Science Education Standards of the National Science Board and the National Council of Teachers of Mathematics (NCTM) Standards, and the reform curriculum is geared to the Florida Sunshine State Standards. The district's core values related to mathematics and science reflect a strong belief in the youth of the county and include the following: (1) all students can learn math and science; (2) teachers must provide maximum opportunities for all students to learn; (3) classroom activities are structured around real-life activities in which math and science concepts are applied; (4) individual and cooperative learning are utilized; (5) collegial interaction is paramount; (6) continuous professional development and growth are expected; and (7) assessment reflects instruction—continuous, performance-based, and authentic.

One program offering, Science Engineering Communications Math Enhancement (SECME), linked students with enrichment opportunities at museums, nature centers, and other community-level resources in the city and county. District personnel reported less well-developed partnership relationships at the school level, but our team of researchers met parents and community residents who were active participants on school governance committees and parent/teacher/student associations at both high schools in the study.

Family involvement was seen by school administrators as attendance at conferences and meetings held at the school. Community or business partnerships were viewed as monetary, physical, or both. Parents themselves had difficulty verbalizing their perceptions of partnership arrangements because many came from cultures that did not encourage parental participation at the school level (e.g., Peru, the Philippines, Haiti). They believed that teachers and principals knew best, and did not feel qualified to give input. This was particularly true with parents of Haitian-descent who had recently arrived or had been in this country for only one or two generations. Definitions of what constituted mobilization of working relationships with parents, community organizations, and other entities that might be considered partners were not uniform throughout the district, and researchers observed that resources also did not seem evenly allocated across all classrooms at the schools in our study. For instance, at "Newcomer" High the new science labs for the Scholars Academy (an advanced placement program) stood in contrast to the science room for ESOL students that did not even contain a working computer or lab table.

Newcomer High School partners expressed the need for more school–community interaction as well as parent involvement, but they had difficulty articulating the nature of that involvement beyond parents taking part in the governance organizations and attending meetings and conferences. Parent/Teacher/Student Association (PTSA) members felt that too much time was spent preparing for the state's standardized assessment (FCAT) at the expense of a more balanced or "well-rounded" curriculum. These same individuals were, however, supportive of assessment as a way to see what and how well the students were learning.

At Renaissance High School, partners thought more support was needed at the district level to implement reform. Miami Beach is generally perceived to be an affluent area; community members thought that the school district assumed the community would take responsibility for students' needs. The high school, a multicultural and diverse institution, attracts students from neighboring communities, not all of them affluent. Most of those we interviewed were not familiar with NSF's reform efforts (USI or USP) or the initiatives' commitment to increasing levels of involvement by school partners, but they were optimistic about the ways the city supported education for all students.

Partnerships and parental involvement were seen as distinctly different in each school and at the district level. District personnel saw partnerships as relationships developed with businesses, community organizations such as museums and nature centers, and service organizations like Chambers of Commerce or Kiwanis clubs. They described parent involvement as a parent-to-school phenomenon involving attending meetings, serving on governance committees, or volunteering in classrooms. Organizations affiliated with schools through district-wide partnerships described themselves as serving *all* students, regardless of school, as long as they met the established guidelines for participation. Students' educational enrichment was their goal, whether through daily support programs, summer camps, tours, speakers, or scholarships. The Miami Museum of Science and the Biscayne Nature Center were examples of this. These large institutions organized regular tours and educational programs for students in the district (some free and some fee-based). Community members in North Miami described their partnerships with schools as a two-way flow. Students, teachers, and student organizations approached organizations for their support (physical or financial), and city organizations approached the school when they needed student participation and volunteers for special events. There were also relationships that brought organizations (or their funds) into the schools. As described by Aber et al. (1997) these kinds of efforts contributed to neighborhood processes to facilitate positive youth development and, in this case, academic achievement. The effectiveness of the partnerships at the school and classroom level was, however, mitigated by forces over which neither the school nor its partners had control such as organizational budget restraints, availability of outlets for volunteering and/or part-time employment for students, and second-language issues (Spanish, Portuguese, or Creole). At the district or school level, little accommodation was made for these and other issues as staff planned strategies for building relationships with parents.

The next section of this chapter provides demographic and historical information about the two communities (table 5.2). This information highlights some of the factors (e.g., income, ethnicity, education) that work to shape the neighborhood and social processes affecting student achievement and implementation of reform efforts. We will then move on to discuss ways relationships are defined by the individuals in the two communities and how those relationships influence reform and subsequent student achievement.

The Communities of North Miami and Miami Beach

Although both Newcomer and Renaissance High Schools are in Dade County, the communities in which they are located vary tremendously with respect to population characteristics of residents, including income, language, and educational attainment. This variation affects the amount of resources that are available to students, parents, and schools to support educational achievement and educational reform. One community's population is predominantly black and Hispanic, with greater numbers of school-aged children and a larger number of people living in poverty. Individuals in this town, almost half of whom are foreign-born, reported West Indian ancestry. The other community, predominantly white (although Hispanics can be included in this category), reported a higher per capita income, fewer families living below the poverty level, and kinship ties with Latin America. In the sections that follow, we will discuss how these differences affect the kinds of relationships that the schools and the school district can develop with the local communities and individuals who lived there.

North Miami FL and Newcomer High School
North Miami, located along Interstate 95, is a largely residential area, part of North Dade County that extends north from Miami to Ft. Lauderdale. Until the late 1970s, when North Miami experienced its first influx of African American homeowners, the community was a predominantly white, low- to middle-income residential neighborhood.

Table 5.2 Demographic information

Category Total population	North Miami 59,880	Miami Beach 87,933
White	20,824 (34.8%)	76,276 (86.7%)
Black (not Hispanic)	32,867 (54.9%)	3,548 (4%)
Hispanic (any race)	13,869 (23.2%)	47,000 (53.4%)
Asian	1,152 (1.9%)	1,202 (1.4%)
American Indian/Alaska Native	191 (0.3%)	206 (0.2%)
Native Hawaiian/Pacific Islander	28 (0.0%)	39 (0.0%)
Other	1,893 (3.2%)	3,557 (4%)
Number of school-age children (5–19 years)	13,908	9,737
Per capita income (1999)	$ 14,581.00	$ 27,853.00
Families below poverty (1999)	2,826 (20.7% of families)	3,165 (17% of families)
Educational attainment (individuals over 25 years of age) high school graduate	67%	78.8%
Nativity—place of birth		
Native born	30,892 (51.5%)	39,209 (44.5%)
Foreign born	29,144 (48.5%) 91% of foreign born from Latin American region, 36% of West Indian-non Hispanic ancestry	48,852 (55.5%) 81.4% of foreign born from Latin American region, 1.4% of West Indian-non Hispanic ancestry
Language spoken at home (> 5 years of age)		
English only	19,252 (34.9%)	27,554 (32.5%)
Spanish	13,501 (24.5%)	46,174 (54.4%)
Other Indo-European language	21,067 (38.2%)	9,169 (10.8%)

In the 1980s, tens of thousands of Haitians migrating to Miami began to occupy an area that is currently known as Little Haiti on NE 2nd Avenue between NE 55th and 70th Streets (Louisna, 2000). The community surrounding Newcomer Senior High, on NE 8th Avenue and 137th Street, is by no means part of Little Haiti, although not far north of it. Haitians moved into North Miami in great numbers throughout the 1980s and 1990s presumably because it is one of Miami's more affordable residential areas (Mardy, 1999). The community simultaneously experienced a wave of "flight" as both Anglo-Americans and American-born blacks fled north and west to Broward county and other areas in South Florida (Vigolucci, 2000). North Miami became and continues to be a predominantly Haitian community, and, despite past tensions, efforts have been made to make this community more active and open to the new residents. One significant example is the night school at Newcomer Senior High. Few Haitians attended the classes offered at the "Newcomer" Adult Educational Center until the first Haitian-American principal in the Miami Dade County Public School System became principal of the center in 1995. Since then, the Center has become one of the best attended of such schools for people seeking to learn English and/or to take high school equivalency or vocational training courses. Enrollment at the night school has doubled in the past five years from 3000 to 6000 students (Mardy, 2000).

Community Ethnography
The team's observations of the community surrounding "Newcomer" Senior High extended to an area included within a one-half to one-mile radius of the school. It should be noted that North Miami has no clear boundaries, but gradually merges with more affluent areas to the south such as Miami Shores and North Miami Beach. The school sits in the middle of a large residential area dotted by small commercial blocks, modest homes, and apartment buildings and some stretches of newer four-story apartment or condo buildings, and bungalows. Houses are generally single-level, small, and old with little yard space and virtually no decorative landscaping. Few new cars were observed in front of the houses; most appeared very old. The occasional house had a carport. These residential areas were, for the most part, deserted from morning until late afternoon, presumably because most adults were working. It was visibly clear that most residents were of African descent, either African American or Afro–West Indian. This aligned with the 200 census statistics cited at the beginning of this section and with

statements of staff at Newcomer High School. Buses ran continuously on busier streets, and a church of some denomination appeared on almost every street corner. The school is located very close to the North Miami Public Library Branch as well as the large Cagni Park and its community center. It was clear that North Miami was, as many informants contended, a working-class community, but one that provided a full range of services for its residents.

At their first visit, the team immediately observed that the school, a faded stucco building, was quite old and somewhat dilapidated. The teachers' parking lot was small and crowded, and there was no student parking lot. Student cars were parked along the median separating the school from the first row of houses in the large residential area adjacent to the school. There were very few new cars, and the number of student cars did not approach the numbers that fill student parking lots of schools in more affluent areas of Dade County. Portables (22 of them in 2000–2001) had eaten up a large section of a soccer playing field as well as a fenced section of pavement that appeared to have once been the student parking lot. The front of the building had been recently painted, and engraved welcome messages in English, Spanish, and Creole were located on a wall next to the large solid doors past the gated entranceway.

Upon entering the building, the researchers were greeted by two green-shirted security personnel who issued guest passes. The hallways of the school were less attractive than the outside of the building, although the corridor in front of the main offices had been recently painted green. Throughout the building chipping red and gray painted walls were being repainted as well. Signs and billboards were present all along the main corridor, many of them in Creole. The bathrooms were stark and dirty, and the walls had clearly been repainted several times. When the bell rang the halls became uncomfortably crowded as students wearing name badges moved from class to class.

An assistant principal provided a list of possible partners and contacts, maps and schedules of classes, as well as room numbers for the chairs of both the math and science departments. The gym, driving range, and cafeteria were old and obviously too small for the number of students in attendance. The school remained active well past the last bell, and students, many wearing sports clothing, came and went until early evening. Although the halls were very crowded, the team saw no conflicts, and every student they approached was relatively polite and helpful.

The researchers attended one PTSA meeting organized to carry out the election of the 2000/2001 officers. Although scheduled to begin at 6:00 p.m., the auditorium remained empty until almost 6:30 when the officers arrived. Only one other parent and three students, one of whose parent was a PTSA officer, showed up by 7:00 p.m. when the meeting began. The team learned that the officers and the parent attending all had children in the International Baccalaureate (IB) magnet program, not in "regular" classes. These individuals spoke to the team about having little support from either parents or the administration, and the low attendance at this meeting seemed to confirm their input. Only one PTSA officer was non-white, and one black parent was in attendance until the night school principal let out a group of parents from classes to attend the meeting. All of these parents appeared to be of Haitian-descent, and they sat together in a small group in the back of the room. It was clear that they spoke little English, and no attempt was made to translate the proceedings. The one black parent who had arrived separately complained that too little notice was given to the parents, and she was clearly displeased with the low attendance at this important election. Through no direct fault of their own, the PTSA officers, their goals and activities, and their perceptions were not representative of the potential body of parents, but they were suffering from the absence of those parents. The IB program had a predominantly white, middle-class student population, and those students' parents had internalized the need to be engaged from their experiences during elementary and/or middle school years. They were not supported at the school level in efforts to reach out to other parents, did not display creative thinking about ways to attract more parents to their meetings, and they were not sensitive to the needs of second-language (Creole or Spanish) parents.

Newcomer Senior High was rated a "D" school under Governor Jeb Bush's "A+ Plan for Education" grading system for the 1998–1999 school year despite its high performing IB magnet program. At the time 3,418 students were enrolled. The student population decreased slightly during the 2000–2001 academic year, when 3,349 students were enrolled. Of this number, 4 percent were classified white non-Hispanic, 2 percent Asian/American Indian/multicultural, 15 percent Hispanic, and 78 percent black non-Hispanic. Although no statistics were available, school officials and partners were certain that over 80 percent of the black non-Hispanic population were actually of Haitian descent. The school's mobility index was 39 percent, and 24.8 percent of students received free or reduced lunch. The percentage of students with limited English Proficiency was 16 percent and the dropout rate was 6.2 percent (Miami Dade County Public Schools, 2002). Tenth grade students who took the FCAT (Florida's high stakes assessment test) scored 266 on reading and 290 on math. The minimum passing score set for test-takers in both categories in 2002 was 300.

Residents' Perceptions about their Community
We asked informants to reflect on their impressions of the assets available to students attending Newcomer High School. Their perceptions about the community surrounding the school were fairly consistent. Respondents thought that most residents and businesses saw the community as a working-class neighborhood, but not necessarily poverty stricken. The housing stock was aging, but because homes were less expensive, it was also a place where families could establish first-home ownership, especially in the Haitian neighborhoods. Businesses were primarily service-oriented—auto repair, tailors, and so on, and the tightly knit Haitian community supported its own businesses, sometimes to the detriment of other businesses. All partners agreed that, with the dramatic increase in the Haitian population starting in the mid-1980s, both the white residential and business communities subsequently declined. Many saw the need for greater school–community interaction as well as parental involvement (in the more traditional definitions of Epstein and others). The latter, most contended, could be fostered by educating the recently arrived immigrants in the Haitian community (from which the majority of African-descent students come) about the more participatory nature of home-school interactions in the United States. This opinion reflected the standard notion of involvement (see Feuerstein, 2000) and was expressed by school staff, members of the Parent/Teacher/Student Association, and staff members of community organizations—representing a range of both incomes and ethnic origins. The informants were empathetic with the myriad of acculturative pressures on recently arrived parents and students, but felt that parents needed to see themselves as welcomed by the schools rather than "giving over" their children to the educational institution. One resource for this effort to modify cultural perceptions about parents' relationship with the schools that was frequently mentioned was the Adult Education Center at the high school. Many parents attended English and vocational classes at the school after working one or more jobs during the day. The principal perceived both Haitian students and their parents to be very committed to successful outcomes, saying:

> The Haitian students in our school are very motivated. Most will be the first person in the family to graduate from high school. Ninety-nine percent of the time, the kid is told by parents that the school is your authority figure during the time your are with them. The parents often

tell us they would like to see corporal punishment reinstated. Our parents are extremely supportive of what we are trying to do. (Misstaff, 2001)

The principal of the Adult School acknowledged that changing the parents' view of the school's role would take time and was not necessarily seen as a priority for many of the parents, but hoped they would build closer relationships with teachers and the school. He was less enthusiastic about urging this same group of immigrant parents to get involved in the governance activities at the school, stressing the survival-oriented activities that took most of their time. Lack of after-school supervision in the home was also an issue, especially as most adults in students' families worked at least one job. The principal stressed the importance of other needs in the community including health care, funding for after-care programs, and accessible vocational or trade schools.

Most partners held similar notions about what they believed should be done to improve conditions at Newcomer High. First, most stressed that a new building should be constructed to replace the existing structure though it was uncertain how this would occur given current district shortages. More computers were a must, especially because many students were unlikely to have computers at home, and the school's library did not have enough computers to efficiently serve all the students and teachers who needed them during the school hours. Many informants believed that some teachers were too comfortable in their positions and had failed to keep current in their own subject matter areas. Hopes for the future tended to be high, and many partners believe that the new principal was instrumental in bringing about important reform and would continue to be a key factor in the future of the high school. They believed that the community cared about the high school, although most did not identify specific agencies. They were unclear about how reform (at any level) could be best implemented and sustained. Some of the informants felt that change would come with or without changes at the district level, while others felt that those reforms driven by MDCPS and the School Board had the support to actually impact educational achievement.

Partnerships and Relationships with the Community
Information about relationships between Newcomer Senior High and entities within the surrounding community was gathered through interviews with two long-tenured teachers, a city hall representative, one police department representative, a parent and PTSA officer, and the director of a local health service organization. Additional data were

gathered by speaking with school officials, including assistant principals, a counselor for college-bound seniors, activities directors, the principal of the adult school, and parents. The school has well-established relationships with both city hall and the North Miami Police Department. City hall offers various programs including essay-writing contests, career mentoring programs with feeder middle schools, and clean-up initiatives in the city. For the past two decades, it has sponsored several students annually to participate in a summer youth development program in Washington. The city's Youth Opportunity Board provides scholarships as well as summer job placement for students. The Chamber of Commerce also provides annual scholarships as well as community service opportunities by enlisting the help of students when it is needed for special events.

The North Miami Police Department works with the school through several programs, the most important of which is the Police Athletic League (PAL), a community-based sports initiative that works primarily through after-school programs. Funding is provided by the city, and funding decisions are made by a police council. The initiative also seeks donations from the community but the response rate is low. The Police Department works with the school through the student-based organization, S.A.D.D., or Students Against Destructive Decisions. Community service is a significant source of school–community interaction, and various clubs within the school are oriented toward this kind of outreach. A new club, SECME, or Science Engineering Communications Math Enhancement, had just started and was intended to support student involvement in the community. Involvement (or service) is necessary for some college-bound students. The Florida Bright Futures Scholarship, a program that guarantees stipends to qualifying resident undergraduates, has a substantial community service requirement. For the most part, existing community service and volunteer efforts are initiated by individual students, a point made by a representative from the American Cancer Society, an organization that provides community service opportunities for students from North Dade high schools. The Kiwanis Club was reported to undertake fundraising for the school. Also noted was the importance of sports in the school, and local businesses were said to provide additional funding and personnel, especially to the football team. There was no specific mention of ways that community organizations support mathematics and science instruction at the school. Other community resources appeared to be established with specific clubs and organizations within the school or with individual students seeking to fulfill their school-mandated community service hours. The ways in which these perceptions impacted the relationships between students, the school, parents and families, and the community will be discussed at the end of this chapter.

Miami Beach FL and "Renaissance" High School
Miami Beach was a deserted, barren, and isolated island until Henry Alum started his coconut farm there in the 1880s. Other farming sites followed, and the island slowly developed. A wooden bridge linking the mainland to Miami Beach was built in 1921. In 1926, Miami Beach was rebuilt in a uniform Art Deco architectural style that distinguishes it today. After World War II, Miami Beach attracted tourists and celebrities, and by the 1960s, condominiums began to sprout to house the growing year-round retirement community. By the 1970s, the community was experiencing a decline. Its buildings were decaying and its population consisted primarily of retirees and senior citizens, earning the city the moniker "God's waiting room" (Moore, 1996). After the Mariel Boat Lift in the 1980s, the Miami Beach Cuban population expanded by the thousands, and tensions followed as the refugees were seen by fearful residents as spurring a crime wave (Kleinberg, 1994). At the same time, Miami Beach experienced an architectural and artistic revival. The Miami Design Preservation League spearheaded the move to get several buildings onto the National Registry of Historic Places. The movement gained national attention in the 1980s with the popularity of the television show "Miami Vice." Preservation efforts continued despite considerable conflicts, and tourists returned to Miami Beach. Today, South Beach, or SoBe, is a mecca of entertainment for residents and tourists alike. It is home to many trendy restaurants, nightclubs, shops, and studios (Moore, 1996). There is another side to this image, however, as is demonstrated by interviews with city staff, observations, and the demographics of the community that show pockets of low income, high mobility, and limited English-speaking households.

Community Ethnography
The beginning of what residents call Mid-Beach sits adjacent to the school in a residential area surrounding a large golf course. The houses were old, many of them quite large and very well kept. Expensive-looking cars were observed in the driveways during working hours, and an occasional boat was parked on the side of a house. The streets—Royal Palm Avenue, Sheridan, and Prairie—were small with minimal traffic. The houses represented a number of architectural styles, and no two homes are alike. The golf course ended at 34th Street, but the residential

area continued to the north and west with gorgeous houses, professional landscaping, and expensive automobiles, especially on the west side, until 40th Street, a major business area with banks, bakeries, and North Beach Elementary School. Several private schools were scattered within the area, most with a religious affiliation. Pine Tree Avenue consisted of more large, ornately fenced homes and tree-lined medians.

Crossing onto the south side of Renaissance High School was like moving onto a different planet. Apartment complexes were suddenly shabby and closely spaced but bordered by expensive condominiums to the east and west. The apartment buildings were small and in need of some painting, but the neighborhood was active and cheery. People of all ages and ethnicities walked their dogs and rode bicycles at all times of the day. Continuing south, the neighborhood looked more and more like the South Beach featured in Hollywood films. Small shops, tattoo parlors, and nightclubs mixed well with the occasional big-name bank or fast-food restaurant. Near the Convention Center and the Gleason Theater—two blocks from the school—was the popular outdoor shopping and dining area, Lincoln Mall. To the east, along Collins and Washington Avenues, the researchers observed many kinds of hotels from run-down to the resort-style tower over occasional cafes and restaurants. South Beach was clearly distinguishable from the residential Mid-Beach, and, although it was more colorful, it was also, quite visibly, much less affluent and undeniably poor in some sectors.

The Miami Beach population is predominantly white (see table 5.2), but this designation does not separate Anglo-Americans from Hispanic Americans who identify themselves as white. At the time of the study, estimates for the poorer, more densely populated areas of North and South Beach were 60–89 percent Hispanic. Mid-Beach, home to most city leaders and an upper-income residential area, was much less densely populated with significantly fewer Hispanics Americans. Income estimates varied substantially across different sections of Miami Beach with Mid-Beach households earning between $39,700 and $300,000, and North and South Beach households earning between $0 and $30,719 (Vigolucci, 2000). Miami Beach is home to both some of the wealthiest and poorest citizens in the Miami area. The Director of Children's Affairs for the city described the community as a three-part phenomenon as well, saying:

> Middle Beach is the affluent, white, upwardly mobile area; . . . South Beach you have of course the entertainment districts, . . . but you also have some . . . residential areas . . . [that are] still affordable housing-type units, for how long I don't know . . . North Beach area, . . . [has] a very high mobility, . . . [and] a lot of very poor housing stock, and

unnecessary sicknesses. Asthma is very high because you have mold and you have leaks and you have lots of other problems. (Miacit224, 2000)

It is clear that this small city prides itself on its progressive approach to serving its citizens, and yet has not escaped the societal problems of poverty and its related pressures on family structures, housing opportunities, and, inevitably, educational achievement for the youth living in the city.

The School
Renaissance Senior High, the only public high school in the City of Miami Beach, is located between South Beach and Mid-Beach, a more stable, upper-income residential area of Miami Beach. The students in 2000–2001 represented a high level of ethnic diversity. Of the 2,652 students, 63 percent were Hispanic, 18 percent were black non-Hispanic, 18 percent classified themselves as white non-Hispanic, and 2 percent reported an Asian/American Indian/multicultural heritage. At that time, the school's mobility index was 39 percent, and 39.3 percent of the students received free or reduced lunch. The percentage of students with limited English proficiency was 20.9 percent, and the dropout rate was 4.3 percent. Renaissance Senior High was also graded as a "D" school under Governor Jeb Bush's "A+ Plan for Education" during the 1998–1999 academic year. A new principal was appointed for the 2000–2001 year. Her mission was to run a tight ship and turn the school in a more rigorously academic direction than that taken by her predecessor. Tenth grade students who took the FCAT in 2002 scored 276 on reading and 300 on math, with the minimum scores set at 300.

In a recent *New York Times* article entitled "E Pluribus Unum at the Dance," a reporter who attended the school's homecoming dance in November 2002 spotlighted Renaissance High School. The principal noted, "the children of multimillionaires come to this school. But we also have students living in cars." The guidance counselor added, "we have Hispanics, African-Americans, South Americans, Europeans, a large Jewish population, a lot of Asians, and even Islamic students." In the article's concluding section, the reporter mused, "less than a century ago Miami Beach was a sleepy backwater where racial restrictions were written into law. Now it is the jittery boomtown that anyone who wants to know the demographic, political, and cultural future of America had better consult" (Trebay, 2002).

Renaissance High's exterior is very attractive, and the school itself is located next to a huge golf course and framed along Dade Boulevard by a well-tended expanse of grass and flowers. The student parking lot was filled to capacity with all types of automobiles including battered Beetles and new BMWs. The building was old but newly painted and clean. Inside, the main office was accessible to visitors and was decorated with a Hall of Fame of notable Renaissance High alumni (white or Hispanic and including actor Andy Garcia). On the office wall a large world map was displayed with tacks indicating every country from which students came. Approaching the main building, the team observed that the path resembled a pier with ropes blocking off a section of aged picnic tables that ended at a guard stand blocking visitor access. Halls were painted pink and were too narrow for the number of students that spilled into them when the bell rang. Outside corridors made the building spacious, and students ate lunch on tables placed along these corridors. Vendor carts sold fast food, ethnic food, and snacks as an alternative to the cafeteria food. Two new science annexes, the only two portables on campus, sat in stark contrast to the main building. They were modern with wide hallways, and the laboratories inside them were large and well equipped. The considerable landscaping and the opulence of the surrounding residential community offset the visible age of the rest of the school.

The 2000 research team visited the school in person to contact the Educational Excellence Student Administrative Committee (EESAC) chair who, on top of his responsibilities on EESAC, was teaching classes and organizing the senior prom. He could not offer the names of any partnering organizations or businesses and suggested that the team speak to particular teachers and school officials. He saw the relationships as supporting individual classes or issues rather than school-wide and did not make reference to district-wide relationships.

In a January 2001 interview, a Renaissance High staff member, newly hired for the pilot full service schools program, identified community-based resources that served students in need of social service support. Twelve entities provided services to students through this pilot program: Hi-Tides Health Center (medical, dental, and mental health services); AYUDA, Inc (a program for at-risk adolescents that offers support and tutorials); Youth Co-op (counseling, academic assistance, summer and school-year employment opportunities); Jewish Family Services TOLL Program (anger management and educational groups); Jewish Family Services Mental Health (off-site provider of counseling services); Family Network (court referral program for students—conflict resolution and parenting groups); A1A Employment (counseling, summer and school-year employment services, postsecondary planning); ASPIRA (dropout prevention program); Barry University (counseling services provided by social work interns); SABER (counseling and academic assistance); Bayview Community Mental Health Center (for students and families); and PAL—Police Athletic League (mentoring and athletics). These services are provided on a contractual basis through the various agencies, but most student-service staff members were based at the school. They received their student referrals either by drop-in or from the guidance counselors. These relationships were all student services-oriented, rather than directly supporting reform or school-wide educational goals. These partners differed markedly from those identified by district-level staff or city representatives. Theirs were agencies or city entities that served the district as a whole or students at the school level (classrooms, volunteer opportunities, etc.).

Residents' Perceptions about their Community
Respondents' perceptions of the school, residents, and businesses in Miami Beach were fairly consistent and emphasized the point that the high school was an important part of the community. Most, however, were aware that many older community members believed the school is not what it used to be either academically or culturally. A principal theme arising from our research indicates that Renaissance Senior was once one of the finest schools in Dade County with astonishingly large numbers (in the 90 percent range) of its graduates going on to college. Talk of disparity between the Renaissance High of the 1930s, 1940s, and 1950s and more recent decades often directed our interviews to the topic of ethnicity. Some school partners talked about affluent Jewish youth who attended the school. The school now has a much broader ethnic base including students from 60 nationalities. Partners explained that changing times and differing cultural beliefs accounted for the disparities; however, the changing economic character of the Beach was mentioned as the most significant factor affecting changes in the school. They described the North Beach section of town as home to a majority of Renaissance High's poorer students.

Gentrification and/or a sharp increase in the price of real estate were topics of conversation for two city employees who worked on educational projects. One of them said:

> I think in order for a community to truly be healthy and to truly have long-term help, both economic as well as community, you have to have a mix. You cannot have all rich or all poor . . . [a]nd one of the things that I see missing . . . each year a little more, is the middle class. I don't necessarily see a defined middle class. And so gentrification . . . if it eliminates this sector of our community is a bad thing, I believe. . . . it's great

for the housing stock but not necessarily for the people living in that housing stock. (Micit211, 2001)

Most individuals noted that they lived in a tourist-oriented community and that few businesses from this sector had established relationships with the school. An exception was a cluster of older businesses on Lincoln Road that hired students to work part-time after school. Parents were particularly concerned that the neighborhood lacked appropriate places for children to be after school. Researchers observed two facilities nearby that offered internships and attractive free-time spaces for students. The first was the Miami Beach Garden Center, a greenhouse, garden, and shop where students interned as cashiers and caretakers. The second was the Miami Beach Community Center where a multifaceted program of both crafts and athletic activities was offered for students. Researchers found that Renaissance High had a remarkable number of community resources such as these that provided opportunities for students to either be employed or be involved as participants.

Interviewees also noted that residents were generally very supportive of the high school, especially those who were parents of high-achieving students. These parents were very active with the Scholars Academy, the advanced placement program at the school, some even participating in curriculum decisions (the EESAC). Parental involvement (in its traditional definition) as a whole, however, was viewed as insufficient as evidenced by poorly attended PTSA meetings. Most individuals stated that more support is needed at the district level to implement schoolwide reform, especially that which included working with parents or the community.

Respondents were moderately optimistic about the future of the school. They had considerable faith in both the new principal and the new PTSA president to enact reforms. Improvements to the building appeared to be on the agenda and were desperately needed. The people interviewed seemed to believe that increased aid would be coming to the school in the next few years from both within and outside of the community. The impetus for these reforms was strong, and they felt that the needed resources would be available. One driving force for change in the public school, however, was the increasing trend for parents with means to place their children in private schools. Although the Scholars Academy is seen in the district as a strong advanced placement program, the school as a whole has a less than stellar academic reputation. One

Miami Beach parent criticized the restrictive nature of the academy curriculum saying,

> The problems deal with the ability for a child to work out a curriculum that includes things that they really want to see in it versus the curriculum that's mandated by the program. (Micit220, 2001)

Parents were choosing to send their children to private schools, depriving Renaissance High of some of the community's high potential students. Some respondents noted, though, that the "average kid" was most neglected. Indeed, as reported in the Florida School Indicators Report for 1998–1999, per-pupil expenditures for "regular" students were $3,816. Exceptional students (including gifted students) received almost twice that with $6,645 spent in that year for each pupil.

Partnerships and Relationships with the Community
Information regarding partnerships was acquired both through visits with school officials and interviews with a former PTSA president, three members of the EESAC, the director of Children's Affairs for the City of Miami Beach, the coordinator for the Scholars Academy at Renaissance High, and three parents. The EESAC, a group that holds monthly meetings at the high school, is made up of teachers, administrators, student leaders, parents, and members of the community. Most parents who attended and the community members who were "Renaissance" High alumni were white and middle class. According to one, the EESAC committee discussed "academic excellence and ways of improving the school" through initiatives they enacted. The conversations often referred to the school of the past, when committee members attended. Other prominent school–community relationships included the Kiwanis Club that in 2000 provided five $2,500 scholarships to Miami-Dade Community College plus $10,000 in additional scholarships for "Renaissance" graduates. One interviewee suggested that Kiwanis was the "largest consistent donor in the school" and had been so for the last 30 years. The Chamber of Commerce was developing a committee to support and fund educational projects and was also a leading supporter of the community education foundation. In addition, the Shane Family Foundation pursued the goal of establishing a school for the performing arts. A key player in the foundation was also the president of Renaissance High's PTSA for 2000–2001.

There is a hospitality magnet program (the Academy of Travel and Tourism) at the school with classes specifically oriented toward careers in the tourism industry. This program allows students to work closely with the hotel association, the cruise line industry, and representatives from tourist businesses on the Beach through mentoring and internships. According to a Miami Beach City Hall representative, government agencies such as HUD, the Department of Juvenile Justice, and the Education Department make partnerships with local education agencies part of their funding requirements now. The school's service clubs, including the Ecology Club, Interact, Key Club, and Future Business Leaders of America, fostered relationships between the school and the community to help meet those requirements.

Respondents insisted that the Miami Beach community was highly supportive of the high school. Interestingly, most partnerships with which they were familiar included those that contributed money rather than time or other assets for the well being of the school. The alumni were visibly and actively involved, and a recently formed Alumni Association was planning the 75th reunion of Renaissance Senior High alums.

One informant, an administrator of the new Department of Children's Affairs for City of Miami Beach saw their partnership as citywide support for education rather than as a traditionally defined partnership with an individual school. She reported that administrators and planners for the city created a Committee for Quality Education in 1999, and she and four other residents worked to create the Miami Beach Education Foundation in 2000. This foundation will, they hope, draw funding to support individual schools or educational programs around the city, independent of district participation. It appeared to the researchers that the city approached education from the top-down, creating vehicles to secure funding to support existing initiatives in individual schools or to create additional educational opportunities for youth in the city. A reciprocal arrangement with Renaissance High School allowed the city to hold its Performing Arts Academy at the school on Saturdays (Miacit224, 2001).

Shifting Definitions of Partnerships and Involvement

Our investigations of two communities and their public high schools revealed that many individuals and organizations undertook types of involvement that went beyond traditional forms of support. Citizens in both Miami Beach and North Miami worked for students from the outside-in, supporting the district through city and county-wide programs (e.g., Miami Museum of Science and Florida International University) and also through city programs that serve larger numbers of students from the immediate area. These programs reach into the schools and make direct contact with classrooms and students to help them achieve in high school. The children's affairs and recreation departments of both cities worked proactively to provide high school students opportunities for personal growth and achievement. A city representative in North Miami talked about her department's relationship with the high school and its individual students.

> We became a Dade Partner officially in 1997 . . . [S]ince the school system requires students to have a certain amount of community service hours and . . . we had the need for about a hundred, a hundred and fifty students to work during the parade, it was a good fit. . . . I would go into the school, have two or three volunteer meetings to recruit students . . . during lunch time so . . . they were able to attend. (Micit221, 2001)

Individual parents, especially in Miami Beach, were involved, not at the school level but at the community level, to build relationships and worked to support students' academic achievement. Because of their personal experiences in the Miami Beach community, these (often) middle-class parents used their personal and social networks as well as their knowledge of resources available to their children to increase students' academic achievement. In North Miami, working-class parents of students at Newcomer High School, often working two or three jobs, deferred their individual contributions to city-wide offerings housed in the Police Department or the Parks and Recreation Department. Students made their own partnerships with these entities, whether for employment, volunteer or service opportunities, internships, or scholarships. Parents become involved at the home level, where they support the choices of their children. As Furstenberg (1999) hypothesizes, the parents with fewer resources rely on home-oriented strategies to support successful student outcomes.

Lower-income earning families in both North Miami and Miami Beach were frequently buffeted by outside forces ("exogenous forces" according to Aber et al., 1997). Such families expected the system, as represented by its institutions including schools, governmental offices, and community organizations, to benefit their children. A major exogenous variable in the Miami case is transnational migration, bringing with it pressures of language difference, low-income levels, and repressive political background experiences that often affected the ways parents and families developed relationships with schools and other institutions to support student achievement. Many adults who have settled in North Miami and Miami Beach came to Florida to escape rapidly deteriorating economic conditions (e.g., Brazil and Argentina) or crushing poverty and autocratic government policies that did little to improve the living conditions of residents (e.g., Haiti and Venezuela). These adults are unaccustomed to having a voice in the day-to-day operations of community institutions and, faced with the realities of surviving in the American economy, focus their efforts on working, learning a different language, and improving their chances for upward social mobility.

Other forces over which both communities and their residents have little control are educational mandates at the federal and state levels such as "No Child Left Behind" that call for "measurable" changes in student outcomes with little acknowledgment of diverse school populations and lived experiences that make uniform progress a difficult goal to achieve. Traditional definitions of partnerships and parent involvement become problematized at the local level because definitions are typically created and sustained by individuals and organizations working from a middle-class bias. Presently, working families (and students) must struggle through changing economic conditions, limited employment opportunities, the new (for immigrants) realities of prejudice, discrimination, and institutionalized racism in the United States.

In this study, both communities were affected by international migration. One of the towns, however, also suffered from the flight of current residents to other towns in the county. The departure of community residents created a loss of stability as individuals and businesses closed and start-ups struggled to take their place in North Miami. The schools in the community lost long-standing allies and, with the onset of an emphasis on academic achievement, turned their attention inward rather than outward to build relationships with the new businesses. The strongest long-standing relationships are between city departments and the schools, and these are sustained by the city, the teachers, and the students rather than by school administrators. Adults in North Miami, with lower per capita income and a higher level of need than was true for those living in Miami Beach, rely on the resources available at the family or household level rather than maximizing support available through institutional resources, supporting claims made by Furstenberg (1999) outlined earlier in the chapter. Students and teachers, however, used institutional resources to meet service requirements for scholarships, secure employment or volunteer opportunities, or gain financial support for their planned activities. Perceptions about the need for more parents to be available to the students were expressed by a city employee:

> [T]he parents aren't involved as they used to be.... [W]e rarely see the parents.... [W]e want to meet the parents and see what's going on in the children's lives.... it's just a sign of the times, I think.... A lot of

them are coming from working parent homes; whereas my mom was always a stay-at-home mom, and it's just changing. (Micit211, 2001)

The situation was different in Miami Beach because a core of individuals who attended schools in the community had now returned with a desire to upgrade educational opportunities. Rather than engage in flight, citizens and institutions in Miami Beach declared they were in the struggle for the long term, creating educational foundations and consolidating institutional support on city-wide committees. There was an acute awareness of the city's economic realities (no longer completely middle class and in agreement about educational goals). Thus, existing institutions developed plans, activities, and services to meet needs of parents and students along a continuum of need. Traditional notions of partnership and involvement were at work in Miami Beach: middle-class parents served on governance and planning committees at the school, and community organizations (e.g., Chamber of Commerce and Kiwanis) supported educational activities at both the school and community level. Parents, many of whom were under the same economic pressures as the parents in North Miami, were absent from the traditional process of involvement as they worked to survive in the face of rising costs in Miami Beach. The economic diversity of the community, however, provided the foundation for a greater level of institutional connections between the schools and students and resources such as libraries, community centers, and volunteer opportunities. Students whose parents were accustomed to reaching out to existing institutional resources were themselves better able to make connections with these resources and secure academic or interpersonal support in order to maximize their chances for successful academic achievement.

In both communities, we also saw the impact of positive relationships on student achievement. This was particularly true in Miami Beach where Hispanic females who made successful connections with community resources also experienced increased levels of achievement in math and science. This finding lends credence to the position that involvement by parents and students is related to improved academic outcomes. Students gain self-confidence, develop networks outside of their family and immediate neighborhood, and are presented with opportunities and options through their connections with city-wide, district-wide, or local school partners. The social and interpersonal processes that support positive student outcomes were at work in both high schools—stimulating students to look beyond their present situations (often poverty and second-language challenges) to see their future potential. These visions supported students' and parents' efforts to achieve positive educational outcomes.

Relationships that parents and community organizations developed with local schools bridged concepts of partnership and involvement. In addition to district-developed school–business partnerships, informal linkages provided individual students with employment and volunteer opportunities. Students accessed institutional resources at the local level to help meet financial or educational goals, to support them during critical after-school periods, and to meet social service needs. Students and partners connected both at the school and community levels, often without formalized agreements between partners and schools. Parents also worked outside traditional roles of parent involvement. They served as employees and volunteers at city-sponsored agencies to meet the developmental needs of all students. Rather than limiting their activities to a particular school, some worked at a city-wide level to establish systems of educational support for all students. These parents, however, were those with more resources—higher-income levels and more leisure time—who could afford to draw upon their experiences and social networks to implement their ideas. Lower-income parents did not, on the whole, engage in school-based relationships, but they supported students' involvement in city-wide or school-sponsored activities to support increased achievement. At both schools, however, students' involvement was attenuated by the necessity for many to supplement the family income by holding part-time jobs after school. Thus, participation in many school and/or city initiatives was strongly influenced by students' family income levels, ability to access transportation, and support from home for this kind of involvement. We saw parents who considered themselves "involved" in their children's education despite their lack of participation in school-based activities. Such involvement generally goes unnoticed by school and district-level personnel; however, it must be a key if schools are to correctly assess the amount (and types) of involvement by community organizations, parents, and students.

In sum, our respondents, newly arrived immigrant families in particular, provided insights into school involvement that expanded researchers' understanding of partnerships and that subsequently resulted in a view of partnerships that also acknowledges the importance of informal relationships with schools and students to support individual achievement. In addition, we learned that system-wide relationships work to help schools and classrooms through student involvement in the work of agencies and cultural institutions as well as businesses. We thus were able to narrow the division between contrasting concepts of "partnerships" on the one hand and "parent involvement" on the other. We found that defining school contacts with individuals or entities as "relationships" allowed us to collapse these two categories and provide a clearer picture of what was happening with and for students at the school and neighborhood levels. Finally, we determined that most relationships had little to do with supporting *specific* reform initiatives (i.e., USI or USP math and science reform), but had a great deal to do with options available to schools, teachers, and students for enhancing students' educational experience.

Notes

1. The authors acknowledge the support of the National Science Foundation through NSF Grant # 9874246: "Assessing the Impact of the National Science Foundation's Urban Systemic Initiative." Any opinions, findings, and conclusions or recommendations expressed in this material are those of the authors and do not necessarily reflect the views of the National Science Foundation.
2. "Drivers" or policy levers were conceptualized by NSF as the framework of influences and outcomes for understanding school reform as a systemic enterprise. Employing an interrelated and overlapping structure of process and outcome reform Drivers (Ds), NSF-funded Urban Systemic Initiative programs throughout the 1990s were designed to be strongly student achievement outcome-oriented (D6, 5), with explicit emphasis on resource convergence (D3), establishing a leadership nexus (D4, 2), partnerships that entail more than the provision of resources (D4), and the advocacy role of community relations including parental and community involvement (D4, 2). The first four Drivers address the creation of a district-level infrastructure that in turn creates a context in which Drivers 5 and 6 can be achieved. Our project aimed at discovering similarities and variations within and across district and school sites associated with both successful and unsuccessful student outcomes as well as the extent to which the NSF formula can be made sustainable.
3. *Methodology.* The studies concentrated on neighborhoods surrounding two senior high schools in each USI site. Each of the schools was assigned a pseudonym. There was no attempt to match or contrast schools by the level of implementation of USI reform; however, selected schools did reflect the diversity of student populations, curricula, socioeconomic status inherent in each site. The methods employed in these studies were useful for understanding contextual influences that are not easily quantified in an evaluation of NSF's Six-Driver model. Interviews, community mapping, and observations allowed researchers to understand the many ways partnerships are defined and mobilized within and across districts.

Using qualitative research methods, researchers conducted individual interviews; teams spent hours in the neighborhoods conducting windshield surveys to identify services, commercial resources, etc.; archival data were retrieved from library, school, and internet resources; and computerized maps were created to highlight pertinent demographic features of neighborhoods surrounding each of the eight high schools in the study.

An interview protocol was developed by the community researchers at a meeting in February, 2000 at the University of South Florida. This protocol (or an adaptation) was used in each site. Researchers interviewed a variety of individuals, conducted walking and driving surveys of neighborhoods around the schools, reviewed archival material in university and public libraries, attended school or community-based meetings, and created maps of the area using information from various local, state, and national sources.

A Miami-based anthropologist pulled the disparate studies together and developed a coding logic useful for analyzing the narrative data derived from the community study interviews. She used Nud*Ist, a qualitative data management program, to identify and assign codes for key elements in the study. The first step in the coding process was to create an Index Tree. A node category was created for each of the four sites. Each document imported into the project was coded in its entirety (at every text unit) for its site. This made it possible for the researchers to look at any text unit and determine from which site it originated. Subnodes of each of these "Site" nodes were also created for demographic information. The inclusion of these subnodes was based on the proposal, which required that researchers include such information in the final site reports. Subnodes were included for categories determined in the proposal, such as heritage, people (economic status and population), and neighborhood descriptions. An "Informants" category was also added to code each document based on the status of the partner-informant.

The remainder of the initial Index Tree codes were based on specific questions asked in the interview protocol, and the organization of the Index Tree followed that of the protocol. If an issue appeared that did not seem to fit neatly into any other node category, then a "free node" was created, which allowed for the inclusion of the information in a way that interacted, but did not cause conflict with, the Index Tree.

References

Aber, J.L., Gephart, M., Brooks-Gunn, J., and Connell, J. (1997). "Development in Context: Implications for Studying Neighborhood Effects." In J. Brooks-Gunn, G. Duncan, and J. Aber (Eds.), *Neighborhood Poverty. Context and Consequences for Children (Volume 1)*. New York: Russell Sage Foundation.

American Automobile Association (2001) Tourbook: Florida. Heathrow FL: Author. 103.

Feuerstein, A. (September 2000). "School Characteristics and Parent Involvement: Influences on Participation in Children's Schools." *The Journal of Educational Research*, 94: 29. Retrieved January 8, 2003 from Expanded Academic Index ASP database.

Furstenberg, F. Jr., Cook, T., Eccles, J., Elder, Jr. G., and Sameroff, A. (1999). *Managing to Make it: Urban Families and Adolescent Success.* Chicago: University of Chicago Press.

Furstenberg, F. and Hughes, M.E. (1995). "Social Capital and Successful Development Among At-Risk Youth." *Journal of Marriage and the Family,* 75(3): 580–593.

Interviews quoted in text of paper. Conducted April 2000, February 2001. Micit 211, Micit 220, Maicit 224, Misstaff2.

Kessler-Sklar, A. and Baker, A. (September 2000). "School District Parent Involvement Policies and Programs." *The Elementary School Journal,* 101: 101. Retrieved January 8, 2003 from Expanded Academic Index ASP database.

Kleinberg, Howard. *Miami Beach: A History.* Miami: Centennial Press, 1994.

Lott, B. (Summer 2001). "Low-Income Parents and the Public Schools." *The Journal of Social Issues,* 57: 247. Retrieved January 8, 2003 from Expanded Academic Index ASP database.

Louisna, G. (February 17, 2000). "Move in 70s Desegregates Neighborhood." *The Miami Herald.*

Mardy, H. (June 8, 2000). "Haitians Answer Principal's Call." *The Miami Herald.*

Miami-Dade County Public Schools. Websites and information retrieved December 29, 2002 and January 4, 2003.
http://www.dadeschools.net/
District performance report for 2000
http://www.dadeschools.net/board/pdfs/SOSA_2002_report1.pdf
School web sites
http://www.dade.k12.fl.us/mbeach/
http://nmhs2.dadeschools.net/index-main.htm
State FCAT Scores
http://www.firn.edu/doe/sas/fcat/fcinfopg.htm
School Profile 2002. Miami: Author.

Moore, M.A. (1996). *Access Miami.* New York: Access Press.

National Science Foundation (2001). *Academic Excellence for Urban Students: Their Accomplishments in Science and Mathematics.* Washington DC: Author

Trebay, Guy (November 10, 2002). "E Pluribus Unum at the Dance." *New York Times,* Sunday Styles Styles Section, p. 1.

United States Bureau of the Census (2000). *2000 U.S. Census Data: American Factfinder Tables for Demographic Information* (http://factfinder.census.gov/servlet/BasicFactsServlet?_lang=en). http://factfinder.census.gov/bf/_lang=en_vt_name=DEC_2000_SF1_U_DP1_geo_id=16000US1245000.html (http://factfinder.census.gov/bf/_lang=en_vt_name=DEC_2000_SF1_U_DP1_geo_id=16000US1245000.html).

United States Department of Education (1998). "Goals 2000: Reforming Education to Improve Student Achievement—April 30, 1998." *Archived Report.* Washington DC: Author.

Vigolucci, A. (January 2, 2000). "Ethnic Make-Over of South Florida Likely to Intensify." *The Miami Herald.*

Part III
School Reform and Public–Private Partnerships

Chapter 6
The Public–Private Nexus in Education

Henry M. Levin

A partnership is generally viewed as a formal agreement between two or more parties that provides mutual benefits to those parties. It is rare that such partnerships exist between public and private elementary or secondary schools. Despite the fact that only about 10 percent of the students are enrolled in private schools, educational institutions in the two sectors are competing for many of the same students and do not find it to their advantage to work together. In higher education there exist a variety of agreements, such as consortia based upon joint sharing of libraries, course registrations, and cooperative programs; however, such partnerships are still modest in scope and are the exception rather than the rule.

At a broader level, there exist many intersections between the public sector in education and various private entities. Whether one would call them policy partnerships is less clear. It is probably fair to say that most formal partnerships between the two sectors are modest in scope. The most prominent of these are public assistance to private schools and business–education partnerships. In these cases, there are formal relationships between government and private schools on one hand, and between private businesses and public schools on the other. In a broader context, it is clear that the education of each child must necessarily be a public–private undertaking to the degree that its success is premised on a parent–school partnership. What students learn depends on not only what happens in school, but also what happens in the home and the degree to which homes and schools are mutually supportive of each other's goals.

In this chapter, I will review a range of linkages between the public and private sectors in elementary and secondary education. I will begin by reviewing the peculiar nature of education in producing what is both a public and private good. This suggests that public–private collaboration should be central to education. I will follow with several existing interventions that link the public and private sectors. I will point out the necessity of public–private collaboration while also stressing the continuing sources of tension between the two sectors when it comes to education. Finally, I will present the most ambitious venture to link public and private sectors in education by providing publicly financed vouchers that could be used for private schools.

Education as a Public and Private Good

Education inherently serves both public and private interests (Levin, 1987). It addresses public interests by preparing the young to assume adult roles in which they can undertake civic responsibilities; values; participate in a democratic polity with a given set of rules; and embrace the economic, political, and social life that constitute the foundation for the nation. All of this is necessary for an effectively functioning democracy, economy, and society. At the same time, education must address the private interests of students and their families by providing development that will enhance individual economic, social, cultural, and political benefits for the individual. Embedded in the same educational experience are outcomes that can contribute to the overall society as well as those that can provide private gains to the individual.

To some degree, the public and private outcomes of schooling can overlap, because better educational results for the individual and her family may also contribute to social benefits. For example, if schooling makes the individual more productive (private benefits), the economy also receives a boost (social benefits). However, in other respects, there may be conflict between public and private benefits. For example, the public benefits of schooling require that students learn to consider different points of view that are presented and debated in the schooling experience. But, the private values may be in conflict with some of these viewpoints, and parents may not wish their children to be exposed to points of view that are at odds with those held by the family.

The problem is that schooling takes place at the intersection of two sets of rights, those of the family and those of society. The first is the right of parents to choose the experiences, influences, and values to which they expose their children, the right to rear their children in the manner that they see fit. The second is the right of a democratic society to use the educational system as a means to reproduce its most essential political, economic, and social institutions through a common schooling experience (Gutmann, 1987). In essence, the challenge in preserving the shared educational experience necessary for establishing a common foundation of knowledge and values that is crucial to reproducing the existing economic, political, and social order (public goals), while allowing some range of choice, (private goals) within that experience. Because the schools represent the primary agency for preparing all students for the major institutions that constitute the bedrock of society, this requirement suggests a schooling process that comprises many common experiences for all students, even if some of these violate the choices that families might make independently for their children.

Both sets of rights are legitimate, and both are partially, but not completely, compatible. It is clear that public schools cannot be advocates for each and all of the many different and incompatible perspectives that parents have regarding culture, language, values, religion, and politics. As a concrete example, there are clearly very strong differences in viewpoint within the polity about the permissibility of allowing abortion. In the larger society, this very controversial matter must be resolved politically. However, emotions run very strongly on the perspectives. On one side of the issue, abortion is considered infanticide. On the other side of the issue, abortion is considered to be a matter of choice for determining the fate of a fetus that is not yet endowed with human properties. This conflict is embedded in deeply held philosophical, religious, and political ideologies.

One group of parents would like the young to see abortion as murder, and another group would like them to see it as family planning, and both sides would consider the contrary view to be illegitimate. This makes it difficult for the schools to present the issue in any form. The easiest route for the schools is to avoid the issue. However, the courts, legislatures, and Congress cannot avoid the issue, and it's one that all citizens should develop an informed understanding that can be communicated through the political process. This is an example of a public dimension of education that may be in conflict with the private goals of many families who want to keep the issue outside of public discourse or to inculcate the correctness of one particular view without debate. Many other public issues cry out for democratic resolution, but families find them objectionable for private reasons. This is often the situation when parents decry the teaching of inappropriate values in public schools because the schools did not inculcate the parents' values. In such a case, parents may put pressures on schools and school boards to make changes, or they may opt to send their children to other schools, public or private (Hirschman, 1970).

Any discussion on public and private issues in education must recognize the tension between public and private benefits and goals. As long as education has both public and private components, there must be a balance and blending. The solution will always be a compromise that will leave some parents dissatisfied. This dissatisfaction will lead parents to pressure schools for change or to escape from public schooling with private schools and home-schooling as possible alternatives. In other cases, parents will move to other jurisdictions that sponsor public schools that are more compatible with their beliefs.

Much of the debate about the proper roles of public schools and the issue of public support for private schools can only be understood within this framework. Parents who believe that schools should be limited to meeting only their private objectives for their offspring will often object to many of the public goals of schooling. Even those parents who accept overall public goals may be at odds with specific activities and goals that are incompatible with their private educational values. Public policy toward education has been to steer a course that embraces the public interest while allowing as much of the private interest as can be accommodated without bringing the two into serious conflict. This is a difficult charge that always places schools under a tension that is not easily resolved. Indeed, Chubb and Moe (1990) have argued that democracy is the problem that besets public education.

Prior to the 1950s, parents and school districts were able to resolve these potential conflicts through what Michael Katz (1971) calls "democratic localism."[1] That is, within each local setting, communities were able to maintain public schools that reflected the predominant politics, values, culture, and wealth. Public schools in much of the nation were segregated by race, and school finance was largely a local matter based upon property taxes that raised more funding for students in wealthier districts, often considerably more, than for those in poorer enclaves. Many children with handicaps were excluded from schools or were provided with inadequate services, and those who were educationally at risk were given no special assistance and were often tracked into dead-end curricula. Inculcation of the religion of the dominant group at the local level was a common feature of school life. Although each of these policies might be incompatible with the public goals of schooling, they were based upon a tacit compromise premised on the view that those with power could influence policies and local practices in directions that would benefit their children over others. Children from groups that lacked powerful advocacy in their behalf, such as African Americans, the poor, and the handicapped were treated in a less enviable way.

But over the next 40 years, decisions and policies set out by courts, legislatures, and Congress reduced these prerogatives and inequalities so that schools became more and more alike, with fewer public alternatives for meeting private educational preferences. Laws were passed that provided special benefits to economically disadvantaged, bilingual, and handicapped students as well as pushed for racial and gender equality. School funding was more nearly equalized between school districts. Official policies of racial segregation were proscribed by law. The press for greater equity removed many of the privileges held traditionally by dominant groups in local communities.

By 1980, a general backlash emerged with the aim of regaining what was lost. If local political power could no longer be used to create schools that echoed the racial preferences, values, religious practices, and wealth of local residents, other alternatives had to be sought. Most of these alternatives revolved around ways to increase local choice within the public schools. Public choice alternatives refer to the ability of families to choose from public schools within a district or from different districts rather than having students assigned to schools (see the essays in Clune and Witte, 1990). Some districts even created magnet schools with special themes to attract families who were interested in those themes (e.g., science, the arts, technology, multiculturalism, business, health professions, and so on). However, even these forms of public choice have been superceded in the last decade of this century by more radical alternatives, such as charter schools and educational vouchers. Charter schools are schools established under public authority that are exempt from many state and local policies and laws as long as they meet the goals set out in their charter (Nathan, 1996). They can be initiated by parents or educators, and they can represent distinct educational philosophies within the broader public context for schooling. Educational vouchers represent the most complete response to the public–private dilemma by funding all schools that meet certain minimal requirements, whether publicly or privately sponsored, with public dollars.

Existing Forms of Public–Private Collaboration

Before addressing the educational voucher initiative, it is important to review briefly existing practices of public funding for private schools, business–school partnerships, and family–school connections.

Public Partnerships

State and local governments provide considerable subsidies to private schools for nonsectarian purposes. The general view is that if state funding benefits the child rather than the religious institutions that sponsor most private schools, it is permissible under federal and state constitutions. Typical government subsidies to private schools are found in the following four areas: tax-free status, textbooks, transportation, and categorical programs.

Because almost all existing private schools are educational institutions that are not for profit, they are exempt from taxes even though they are eligible for all pertinent local and state services supported by tax revenues. Textbooks that are provided to public schools for the standard nonreligious courses are often provided free of charge to private schools. Many states also provide transportation of students to private schools on the same basis as public school students. Finally, federal and state programs for disadvantaged students are often offered at private schools in classrooms that are not adorned with religious symbols and staffed by employees from the local public school district. One early study (Sullivan, 1974) found that about one quarter of the cost of private schools is borne by government, but that study was done some two decades ago when the law was more restrictive, so the portion is likely to be much higher now.

Business and Families

Businesses have had a long tradition of establishing partnerships with schools in a variety of ways. Usually these are based upon both self-interest and altruism. Such partnerships can improve the preparation of the labor force hired by businesses and provide good public relations, but they can also be forged in the spirit of community involvement. The forms of such partnerships are widely varied. At the local level, they include adopt-a-school programs that offer financial assistance to schools, expertise in particular subjects or managerial challenges, release time to employees for being volunteer tutors, and awards for student performance. At regional and national levels, they may include formation of private associations to provide political support for school reform as well as larger grant programs that assist schools to make major changes. For example, IBM sponsors grant competitions with awards in the millions of dollars to school systems that will make a significant commitment to new applications of computers and related technologies.

Schools have also had a long tradition of cooperative work arrangements with businesses for training and placing students in vocational studies (Steinberg, 1997). By providing part-time jobs for students that relate to their vocational preparation, these businesses offer both applied experiences and income. Such business arrangements may also include gifts of equipment and funds to support vocational programs. In addition, the same businesses will often hire the graduates of these programs if they have performed satisfactorily. This is an example of a mutually beneficial partnership activity because it supports both the learning and training of a local workforce.

Schools and Families
At a less formal level, but even more consequential in terms of student outcomes, there is a tacit partnership between public entities (schools) and private ones (families). It is well known that student achievement in schools is heavily dependent upon family influences. In particular, children from families of higher socioeconomic origins, with higher income and parental education, tend to have better educational achievement than those from lower, more modest origins (Natriello, McDill, and Pallas, 1990). The former type of families is better able to provide the resources and experiences that support school learning.

It is useful to separate family influences on learning into two parts. The first part consists of the natural interactions that more educated and affluent families have with their children that lead to educational success. Such families use a standard version of the English language, an educated vocabulary, and styles of interaction that tend to be more oriented toward questioning and reasoning techniques (Heath, 1983). These are the types of interactions that lead to the knowledge and behavior that schools build upon and achievement tests measure. In addition, their higher incomes mean that students are exposed to a richer set of worldly experiences that contribute to their education. Examples of this range of experiences include travel, computers, summer camps, books, hobbies, and music lessons. Finally, they are more able to provide nutrition, health, counseling, tutoring, and other inputs that support school learning. However, in addition to these, there are specific practices that families can engage in, with respect to the schools their children attend, that will improve both their children's chances of success and the quality of the schools.

Joyce Epstein, the foremost scholar in the area of school, family, and community partnerships, has identified six types of family involvement (Epstein et al., 1997).

1. Parenting—helping families establish home environments to support children as students.
2. Communicating—designing effective forms of school-to-home and home-to-school communication about school programs and student progress.
3. Volunteering—recruiting and organizing parental help and support for school.
4. Learning at home—assisting families to help students at home with homework and other school-related activities.
5. Decision making—including families in school decisions and developing parent leadership.
6. Collaboration with community—using community resources to support families, strengthen schools, and increase student learning.

It is noteworthy that these activities represent forms of school support for families and communities, and forms of community and family support for schools, both efforts focusing primarily on improving student success. An excellent handbook for action in all of these areas is found in Epstein et al. (1997).

Educational Vouchers

Although the concept of educational vouchers has been around for at least two centuries, the specific form that has been debated in recent years dates to an important publication by Milton Friedman on the role of the state in education (Friedman, 1962). In that work, Friedman argued that schools should be funded by the government because of their importance in producing the values required for democratic functioning. Although Friedman called these "neighborhood benefits" they are similar to what we have referred to as the public benefits of education, contributions to the larger society rather than just the individual. Friedman argued that just because government finances schools, it does not mean that government should operate them. Suggesting that the government was an unresponsive monopoly, he asserted that schools ought to be placed in the competitive marketplace that would promote a plethora of for-profit and not-for-profit schools. To accomplish this, the financing of schools would take place through government-issued vouchers that could be applied toward tuition at approved schools that met minimal requirements for assuring the public interest. These vouchers would be redeemed at the state treasury by schools, and parents could add on to the vouchers if they had the means and the commitment to do so.

According to Friedman, such a plan would assure efficiency, innovation, and responsiveness to parental concerns through the incentives of the competitive marketplace. Schools would emerge to serve particular market niches and compete between themselves, and parents could shift their patronage from schools that displeased them to ones that are more attractive. Furthermore, a much larger variety of schools would arise to serve the private interests of families, while protecting the public interest through minimal regulations on curriculum. Thus, the Friedman proposal acknowledged the existence of both the public and private benefits of education while creating a financial mechanism for the private marketplace that would presumably allow attention to both.

Whether the voucher plan that was proposed by Friedman would do all that he claimed has been a source of contention ever since. Friedman's initial voucher plan was shy on details with respect to the size of the voucher; regulations that would assure the production of public benefits; and provision of information to both schools and prospective producers, on one hand, and families, on the other, a prerequisite for a competitive market. Thus, a number of different voucher plans have arisen over the years that have made concrete provisions in each of these areas with somewhat different goals for each plan. Among the most notable are the plans proposed by the Center for the Study of Public Policy (1970) that were designed for a voucher experiment to be administered by the Office of Economic Opportunity as part of the Poverty Program; the proposal for transforming state school systems to vouchers by Coons and Sugarman (1978); and the plan by Chubb and Moe (1990), which caught the attention of many school reformers in recent years. In addition, publicly sponsored voucher demonstrations have been taking place in both Cleveland and Milwaukee, and privately financed voucher projects have been sponsored in San Antonio, New York, and Indianapolis (Moe, 1995).

Differences between voucher plans can be understood largely in terms of three dimensions: finance, regulation, and information (Levin, 1991).

Finance
Central to the potential impact of vouchers on equity is the size of the voucher and the issue of whether families can add their own resources to school payments. Friedman's original voucher plan would suggest a flat voucher of modest value with parental add-ons to that voucher if the parents had the means and desire. Later voucher plans typically limit

parental add-ons and include compensatory vouchers, such as larger vouchers for the poor and the handicapped to compensate for the higher costs of meeting their educational needs (e.g., Center for the Study of Public Policy, 1970). In addition, school participation requires financial provision for transportation so that the many parents who cannot provide this for their children because of costs or work schedules can gain access to potential alternatives. The initial Friedman plan does not discuss transportation, but it is recognized as a requirement by later plans.

Regulation
Even Friedman suggests that voucher schools should be subject to some curriculum regulations to ensure that they produce public benefits, although such regulation would be minimal. However, subsequent voucher plans such as that of the Center for the Study of Public Policy (1970) or Coons and Sugarman (1978) would require a variety of other measures, including regular reporting of achievement test results of their students. In addition, they would require nondiscrimination in admissions and a lottery for some portion of their admissions if a school received more applications than it could enroll. Stringent curriculum and teacher licensing requirements have also been debated as requirements for schools to be approved to redeem vouchers.

Information
Efficiency in competitive markets requires that substantial information be available to both buyers and sellers. For example, families need to know the available alternatives and their educational consequences. Although the Friedman plan makes no provision for gathering and disseminating information on schools, other plans typically assume some responsibility for doing so.

In summary, there is no single voucher plan, but many different ones with different provisions that auger for different educational outcomes. Some tend to focus more fully on maximizing family choice, whereas others would sacrifice some choice through funding and regulations that would emphasize equity and a common core of learning.

Four Major Dimensions

In order to understand the arguments for and against educational vouchers and public dollars for private schools, it is important to identify four major criteria that emerge in the public debate. Each of these following criteria is highly important to particular policy makers and stakeholders: freedom to choose, efficiency, equity, and social cohesion.

Freedom to Choose
For many advocates of vouchers, the freedom to choose the kind of school that emulates their values, educational philosophies, religious teachings, and political outlooks is the most important issue in calling for educational change. This criterion places a heavy emphasis on the private benefits of education and the liberty to choose schools that are consistent with the childrearing practices of families.

Efficiency
Perhaps the most common claim for educational vouchers is that they will improve the efficiency of the schooling system by producing better educational results for any given outlay of resources. Numerous studies have been done that attempt to measure differences in student achievement between public and private schools or between students using vouchers in private schools and similar students in public schools in the few cases of voucher demonstrations (Levin, 1998; Metcalf et al., 1998; Peterson et al., 1998).

Equity
A major claim of those who challenge vouchers is that they will create greater inequity in the distribution of educational resources and opportunities that may result from gender, social class, race, language origins, and geographical location of students. Voucher advocates argue that, to the contrary, the ability to choose schools will open up possibilities for students who are locked into inferior neighborhood schools, and the competitive marketplace will have great incentives to meet the needs of all students more fully than existing schools.

Social Cohesion
As set out earlier, a major public purpose of schooling is to provide a common educational experience with respect to curriculum, values, goals, language, and political socialization so that students from many different backgrounds will accept and support a common set of social, political, and economic institutions. The challenge is whether a marketplace of schools competing primarily on the basis of meeting the private

goals of parents and students will coalesce around a common set of social, political, and economic principles in the absence of extensive regulations or powerful social incentives.

Evidence

The desirability of a voucher approach will depend upon how effective educational vouchers are relative to the existing alternatives on each of the four criteria as well as how much weight is attached to each criterion. It is important to note that if a particular dimension is not valued highly by a constituency, the evidence will not matter very much for that dimension. That is, preference for vouchers or a particular voucher plan is not completely dependent upon evidence on all of its dimensions, but only on what is deemed important by the observer. The fact that no full-fledged voucher program has been tested in the United States means that evidence is limited. However, in the 1990s there has been a considerable outpouring of empirical literature on some of the voucher demonstrations, differences in achievement between public and private schools, studies of choice patterns, and costs that can be used to partially examine these issues (a summary is found in Levin, 1998). On the basis of the literature, as well as the overall knowledge of how markets function, some conclusions might be drawn. However, even these conclusions will depend ultimately on the type of voucher plan that is being considered. For example, voucher plans with minimal regulations may have very different consequences from ones that are highly regulated.

With respect to the criterion of freedom of choice, the voucher alternative would seem to be superior in giving families a wider variety of possibilities that might match more closely their private goals in raising their children. The gap in favor of educational vouchers would be widest when compared with a traditional school system in which children must attend their neighborhood schools. The gap will narrow in those cases where public schools include intra-district and interdistrict choices and magnet schools, and it will be narrowest when charter schools, with their quasi-independence, are allowed. Obviously, freedom of choice will depend heavily on the existence of and access to alternatives, factors that are dependent on the provision of transportation and good information.

With respect to the efficiency of schools under educational vouchers, we can divide the phenomenon into two types, micro and macro. Microefficiency refers to the ability to maximize educational results at the school site. Obviously, if different schools are producing different types of educational outcomes to please their clients, comparisons will be difficult. Indeed, market advocates would view the fact that parents could choose the kind of education that they want for their children as a major dimension of using resources more efficiently. Voucher detractors would argue that the absence of the public goods aspect of education in the market solution means that the voucher schools simply produce more of the private benefits at the expense of public ones.

When student achievement is used as the measure of educational result, it appears that private schools and those under voucher arrangements might have a small advantage over public schools with comparable students (Levin, 1998), possibly because they are able to more readily focus on a narrower range of outcomes than those under democratic control (Chubb and Moe, 1990). Studies examining differences in educational achievement between students in public and private schools or in voucher demonstration projects show a private school advantage, although the differences are small (Levin, 1998; Metcalf et al., 1998; Peterson et al., 1998). Typical differences are a few percentiles and are limited to one or two subjects out of four or five that have been measured. For example, after two years, the Cleveland voucher demonstration found advantages for voucher students over comparable public school students in language, but not in reading, science, mathematics, or social studies (Metcalf et al., 1998).

Macroefficiency includes not only results at school sites, but also the comparative costs of the overall infrastructure to maintain an educational voucher system relative to the overall costs for maintaining the existing system. Particular areas of such infrastructure include record keeping, school accreditation, transportation, information, and adjudication of disputes. Clearly, some of the costs of a voucher system will depend upon the provisions that are put in place. For example, if microefficiency benefits are to be obtained through competition, then a substantial investment in information and transportation may be required. If schools are to be accredited for vouchers on the basis of meeting the requirements for producing public benefits, a monitoring agency will be required. Even in the absence of these provisions, the cost of record keeping will rise as a central agency must keep track of student attendance, voucher eligibility, and redemption of vouchers on a statewide basis.[2] A study by Levin and Driver (1996, 1997) makes a first attempt at reviewing these measures of supportive infrastructure and finds that such costs for a system of educational vouchers would be considerably higher than for the existing system.

In summary, educational vouchers would promote higher efficiency at the school site, but the costs of infrastructure to support such a system would be considerably higher than that of the present system. On balance, it is difficult to say whether macroefficiency favors one system or the other in the absence of greater detail about the features of the voucher system and the setting where it would be emplaced. Furthermore, without taking account of the consumer gains from freedom of choice and the potential losses of public benefits, it is not clear which approach is more efficient in the use of resources.

Although the existing system of public schools is highly stratified by race and social class, as well as fiscal inequities, most analyses of educational vouchers suggest that they would increase inequities. There are three reasons for this conclusion. First, any voucher plan that allowed add-ons to the government-provided voucher would favor families with higher incomes and fewer children. A lack of investment in both transportation and an effective information system would also favor those who are better off because of their abilities to afford transportation and access information. Second, the evidence from many studies on educational choice finds that the poor are least likely to take advantage of choice, and that both family selection and school selection lead to "cream skimming."

The first of these could be countered by specific provisions that favor the poor, such as compensatory vouchers that are larger, transportation solutions, and effective information strategies. Whether these would be adequate to reduce inequities relative to the existing schools is not clear, and the costs of infrastructure to support a more equitable system would be high.

Finally, the criterion of social cohesion is the one that would seem to be more conducive under public school systems than an educational marketplace. The very appeal of freedom of choice is to send children to schools that emulate the specific values and goals of individual families rather than the common goals of society. Schools would rise up to compete for specific market niches by religion, political orientations, national origin, language, culture, and other salient dimensions. The common values and institutions that are required for addressing public goals of education would be undermined by such market behavior. Only through heavy regulation, which inhibits freedom of choice, can attempts be made to coerce schools into producing these public benefits.

The Voucher Debate

Those who believe that the issue of vouchers will be resolved by a spirited search for empirical evidence on some of these dimensions may be severely disappointed. Much of the support for or opposition to educational vouchers is premised on ideology and values rather than evidence. For those who believe strongly in freedom of choice in schooling and maximization of family preferences, the issues of equity and social cohesion may not be important, regardless of empirical findings in these domains. For those who believe strongly in social cohesion and equity, the issues of family preference and choice may not weigh heavily. Indeed, this seems to be why both sides have tended to limit the debate largely to efficiency and effectiveness comparisons of public schools with private and voucher schools, a matter that both sides agree has some importance. Ultimately, the matter will be decided more on the basis of values and political might than on evidence of which is superior. And the struggle between those who view schools predominantly for their private benefits and those who view schools predominantly for their public benefits will continue to challenge and modify whatever system is put into place (Camoy and Levin, 1985).

Notes

1. The argument in this section is developed more fully in Levin (1987).
2. For example, in California, a state agency would need to shift from keeping track of about 1,000 school districts to maintaining records on about 6,000,000 students and as many as 25,000 schools. See Levin, 1998.

References

Camoy, M. and Levin, H.M. (1985). *Schooling and Work in the Democratic State*. Stanford, CA: Stanford University Press.
Center for the Study of Public Policy (1970). *Education Vouchers, a Report on Financing Elementary Education by Grants to Parents*. Cambridge, MA: Author.
Chubb, J. and Moe, T. (1990). *Politics, Markets, and America's Schools*. Washington, DC: The Brookings Institution.
Clune, W. and Witte, J. (Eds.) (1990). *Choice and Control in American Education*. New York: Falmer Press.
Coons, J.E. and Sugarman, S. (1978). *Education by Choice*. Berkeley, CA: University of California Press.
Epstein, J.L., Coates, L., Salinas, K.C., Sanders, M.G., and Simon, B.S. (1997). *School, Family, and Community Partnerships*. Thousand Oaks, CA: Corwin Press.
Friedman, M. (1962). "The Role of Government in Education." In M. Friedman (Ed.), *Capitalism and Freedom* (chap. VI). Chicago: University of Chicago Press.

Gutmann, A. (1987). *Democratic Education*. Princeton, NJ: Princeton University Press.
Heath, S.B. (1983). *Ways with Words*. New York: Cambridge University Press.
Hirschman, A. (1970). *Exit, Voice, and Loyalty*. Cambridge, MA: Harvard University Press.
Katz, M. (1971). *Class, Bureaucracy and Schools: The American Illusion of Educational Change*. New York: Praeger.
Levin, H.M. (1987). "Education as a Public and Private Good." *Journal of Policy Analysis and Management*, 6, 628–641.
Levin, H.M. (1991). "The Economics of Educational Choice." *Economics of Education Review*, 10, 137–158.
Levin, H.M. (1998). "Educational Vouchers: Effectiveness, Choice, and Costs." *Journal of Policy Analysis and Management*, 17, 373–392.
Levin, H.M. and Driver, C.E. (1996). "Estimating the Costs of an Educational Voucher System." In W.J. Fowler, Jr. (Ed.), *Selected Papers in School Finance* (NCES 96–068). Washington, DC: Department of Education, National Center for Educational Statistics.
Levin, H.M. and Driver, C.E. (1997). "Costs of an Educational Voucher System." *Educational Economics*, 5, 265–283.
Metcalf, K.K., Muller, P., Boone, W., Tait, P., Stage, F., and Stacey, N. (1998). *Evaluation of the Cleveland Scholarship Program: Second-Year Report, 1997–98*. Bloomington: The Indiana Center for Evaluation, Indiana University.
Moe, T.M. (1995). *Private Vouchers*. Stanford, CA: Hoover Institution Press.
Nathan, J. (1996). *Charter Schools: Creating Hope and Opportunity for American Education*. San Francisco: Jossey-Bass.
Natriello, G., McDill, E., and Pallas, A. (1990). *Schooling Disadvantaged Children*. New York: Teachers College Press.
Peterson, P.E., Myers, D., and Howell, W.G. (1998). *An Evaluation of the New York City School Choice Scholarships Program: The First Year*. Washington, DC: Mathematica Policy Research.
Steinberg, A. (1997). *Real Learning, Real Work*. New York: Routledge.
Sullivan, D.J. (1974). *Public Aid to Nonpublic Schools*. Lexington, MA: Lexington Books.

CHAPTER 7
GOVERNANCE AND ACCOUNTABILITY
IN THE MICHIGAN PARTNERSHIP FOR
NEW EDUCATION: RECONSTRUCTING
DEMOCRATIC PARTICIPATION

Lynn Fendler

From 1989 to 1996, the Michigan Partnership for New Education (MPNE) operated as an institution that joined Michigan State University (MSU), public schools, businesses, local governments, and communities into a cooperative network. Organized around the purpose "to improve the educational outcomes for Michigan's children," the Partnership offered advantages to all participants in exchange for their contributions. Local schools received money, resources, and expertise; the university was provided with research venues and student–teacher placements; communities and businesses gained a formal voice in educational policy-making; and everyone was offered the prospect of a better-educated workforce. In these ways, the Partnership promised to guide school reform in a way that would benefit a wide array of constituents in Michigan. The keystone of the Michigan Partnership was the establishment of Professional Development Schools (PDSs), in which the Teacher Education Department of Michigan State University worked closely with local schools to provide professional development and school improvement. In August 1993, 26 PDSs were active in Michigan.

The Partnership operated during a time when popular sentiment about educational reform ran high. The Michigan Partnership was a formidable fiscal and bureaucratic entity in its own right, but it also had significant ties to the Holmes Group (1987–1997).[1] Both the Michigan Partnership and the Holmes Group were affiliated with MSU, and some of the same people participated in all these institutions.[2] Especially during the early years, the Partnership operated with some degree of public visibility, especially since the Board of Directors of the Partnership included prominent political and corporate figures.[3] With an opening budget of $48 million, the Michigan Partnership was the largest public–private education partnership in the United States (Bradley, 1990).

The rhetoric used by the Partnership conveyed a missionary zeal:

> Although the will to improve is essential, commitment alone is not enough to effect the revolution that is required in classrooms across the state.... New means must be found for connecting educators with families, and schools with communities, on behalf of powerful learning for all our young people. (Michigan Partnership for New Education, 1993–1994, p. 2)

Partnership discourse defined the purpose of education as preparing people for successful competition in a global economy:

> Global competition is fundamentally transforming work in America. To respond, we need workers who can *take charge* of their work and their lives, *connect* and work effectively with widening circles of others, and *command* discipline-based knowledge and know-how. These same qualities are required to meet the demands of citizenship and to lead a satisfying individual life. (Michigan Partnership for New Education, 1993–1994 Plan, p. 1; emphases in original)

This chapter studies events in the Michigan Partnership for New Education between 1993 and 1995 as a particular case of governance in educational partnerships. This critical analysis of the statements, goals, and configurations that constituted the Michigan Partnership provides a way to understand how a partnership responds to and contributes to shifting relations of participation among public, private, state, and business sectors. Using planning and evaluation reports from the Michigan Partnership as sources, this chapter examines how partnership discourse constructed notions of governance and accountability in a wave of educational reform. I argue that the articulation of alliances appeared to include a broad-based coalition of stakeholders. However, and at the same time, the Michigan Partnership discourse inscribed a historically specific form of school governance and accountability. The analysis of the Michigan Partnership in this chapter is a critical history of a case that contributes to an understanding of how an educational partnership can construct new roles and meanings for civic participation. The shifting meanings of governance and accountability are constructed around two pivotal tensions: centralization versus decentralization, and product- versus process-driven approaches of governance. Understanding the shifts in these sets of relations has bearing on the direction of continued reform efforts for U.S. schools.

Governance in Partnership Discourse: Reconfiguring Centralization and Decentralization

Taken together, various recent educational reforms appear to pull in opposite directions. Going in one direction, charter schools, voucher plans, site-based management, and home-schooling seem to be decentralizing or centrifugal movements. Going in the other direction, reforms like state-mandated curricula, standardized testing, state-distributed funding, and top-down takeovers of schools appear to be moving in a centralizing or centripetal direction. Some educational literature characterizes these movements as indications of competing ideologies, however, other analysts have argued compellingly that these various reform tendencies are neither oppositional nor unidirectional (see e.g., Popkewitz, 1998; Mintrom, 2001). In fact, these various reforms can be seen as complementary efforts that combine to construct a historically specific constellation of power relations. When these reform movements are examined in their historical circumstances and in relation to one another, it becomes more difficult to draw sharp analytic distinctions between centralizing and decentralizing reform efforts. Here I focus on three realms—governing bodies, funding patterns, and school types—that illustrate how centralization and decentralization practices interweave to construct a partnership of particular social fabrications.

Governing Bodies in the Third Way
The political ethos in the United States, at least since the Progressive era, can be characterized by two complementary impulses. A centralist impulse dances with a decentralist impulse in a political *pas de deux* that expresses a specific construction of social individuals in the United States. Similarly, expert knowledge and popular opinion together constitute the voice of reason in educational decision-making. Some historians have analyzed the relations among these impulses as power struggles that have shaped educational policy and curriculum since the advent of common schools in the nineteenth century (Kliebard, 1986; Tyack, 1974). However, it is also possible to view these impulses in their mutual and complementary relations in order to examine the consequences of their combined effects.

School reform efforts are historically situated. This means that the efforts to reform schools are shaped by the historical ethos of the moment; simultaneously, school reforms comprise one dimension of any historical ethos, so the relationship of school reform to historical trends is a mutually constitutive one. Recently, academic and popular discourse has begun to popularize a term that crystallizes a long-standing and broad social movement that aims to avoid political extremes. Tony Blair, Anthony Giddens, and Amitai Etzioni have all popularized the term "Third Way" as convenient language in which to talk about synthesizing the two unsatisfactory options of state control and free-market individualism (or, in political science terms, communitarianism and liberalism). Anthony Giddens (1998) helped to codify the language of "Third Way" when he wrote, "having abandoned collectivism, third way politics looks for a new relationship between the individual and the community, a redefinition of rights and obligations" (p. 65). Giddens uses the terminology of third way to advocate "democratizing democracy," and his thinking resonates with the rhetorical appeal of school partnerships: "Government can act in partnership with agencies in civil society to foster community renewal and development" (p. 69).

In a similar vein, Amitai Etzioni (2000) found the rhetoric of the third way useful in United States contexts when he wrote:

> It is correctly stated as neither a road paved by statist socialism nor one underpinned by the neoliberalism of the free market. It tilts neither to the right or [sic] to the left. (In the US—which has had no significant social tradition—the Third Way runs between a New Deal conception of the big state, which administers large-scale social programmes and extensively regulates the economy, and a libertarian or laissez faire unfettered market.) (Etzioni, 2000, pp. 13–14)

The discourse of the Michigan Partnership is a case of school reform jumping on to the bandwagon of Third Way thinking, and at the same time, the reforms generated by the Partnership serve to naturalize the acceptability of Third Way thinking. Two organizational technologies of the Michigan Partnership serve to fortify the appeal of Third Way thinking as they capitalize on the global popularity of Third Way rhetoric. I discuss first the mobilization of public opinion polls, and second, a "management" approach to school governance.

In 1982, the Michigan Department of Education (MDE) created Project Outreach, which conducted yearly public opinion polls to monitor attitudes about education. Beginning in 1991, however, the polling work was contracted to a privately owned public relations think tank called Public Sector Consultants. Shifting the survey work to a private company gave the appearance of government accountability insofar as the survey appeared to be a nonpartisan, research-based evaluation of the Department of Education. Moreover, the Public Sector Consultants was a professional data-generating entity whose findings could be regarded as authoritative regarding public opinion. In this way, supporting data were generated in a way that represented partnership sensibilities as validated by the currency of Third Way thinking; opinions were solicited from public and private constituents.

The results of the 1993–1994 poll were compiled and presented in a Michigan Education report written by William Sederburg, vice-president for Public Policy and Director, Public Opinion Research Institute. The opening paragraphs of that report are indicative of the alphabet-soup network of government, university, civic, and private entities that were invested in the design and execution of the public opinion poll:

> This is the third year that private sponsors have underwritten the survey. The commitment of the Michigan Business Leaders for Education Excellence (MBLEE) and the Michigan Partnership for New Education (MPNE) to improving education enabled Public Sector Consultants (PSC) to conduct the 1993–94 survey [sic] in cooperation with the Michigan Department of Education.
>
> The Survey Instrument was developed by PSC and the MDE with significant input from a number of parties. Special thanks are extended to Dr. James Phelps of the MDE, who contributed the basic research design; Tom Vance, public relations specialist from The Upjohn Company (representing MBLEE); Henrietta Barnes of the MPNE; Jim Sandy, executive director of the MBLEE; Dr. Georgia VanAdenstine, Governor John Engler's education advisor; Dr. Pat richie [sic] of the Michigan Education Association; and a number of legislative staff members who contributed ideas and questions. Data from the survey are public information. (Sederburg, 1994)

The Michigan Business Leaders for Education Excellence consisted primarily of business sector interests; the MPNE was a conglomeration of university, business, private, and governmental sectors. The Michigan Education Association is the major teacher union in Michigan, and the governor's office had its own representative in addition to the official participation of the Michigan Department of Education. Appealing to Third Way rhetoric as a way to sound reasonable, the polling sample was assembled deliberately from populist and academic sectors. This provided the Michigan Partnership with strong rhetorical support for soliciting foundational support and influencing public policy.

The formally designed public opinion survey fabricated a voice of authority in Michigan Partnership debates. Public opinion polling is not a new mechanism in political machinations, however the Third Way had become an effective rhetorical tool that made it seem as though this voice represented neither the state nor the university. The poll appeared to be both inclusive and non-biased. The shift in the construction of civic participation here is two-fold. First, entities like Public Sector Consultants function like disinterested reporting media; such consulting bodies have no traditional affiliation with state governments, universities, or businesses, therefore, they appear apolitical and unaligned, a stance that had been made to seem very desirable in the fashion of the Third Way. The work of Public Sector Consultants is virtually invisible. Public opinion is reported in a way that makes surveys appear neutral and categories objective, even though we all know they are not. The report is ostensibly authored by a faceless corporation, not by a person; its particular biases are disguised as nonpartisan election results. This way of producing knowledge leads to the second shift in the construction of civic participation. Public opinion polls give the impression that they already encompass dissenting points of view. This is not radically different from elections that are mediated by national committees and advertising corporations; however, the specific mechanisms of spin and image creation are different because focus groups are not run like elections. Opinion polls take the place of referenda, and focus groups are the authorized forms of civic debate. Then, in the reporting phase, disagreement and critique are normalized as survey results produce public knowledge about "majority opinion."

By diminishing the perceived need for debate, the public opinion poll functions as a social technology that produces a docile citizenry (see, e.g., Foucault, 1979). The pretense of public exchange of dissenting points of view, as an ideal of participatory democracy, has shifted to an ideal of focus-group representation through expert polling consultants. Citizenship responsibilities now entail the willingness to respond to polling surveys, participation in focus groups, and acceptance of reported results as representative of popular opinion.

In addition to the public opinion polls, another discursive move capitalized on the fashionable appeal of Third Way thinking. Michigan Partnership documents began to use the term "management approach" to apply to school governance. *The 1993–1994 Michigan Education Poll: Focus on Reform* generated an analysis that explicitly identified a political opposition between centralist and decentralist reform agendas. The report characterized the centralization position as the "systems approach" (endorsed by the teachers' union), and the decentralization position as the "choice/competition approach" (endorsed by Governor Engler).[4] Significantly, the 1994 report of public opinion went on to suggest a third approach, which they called the "management approach":

> The *management approach*, which emphasized school management as a means to improve schools, *did not receive much attention* during the late 1993 debate but has reemerged in 1994. Instead of strengthening the components of the education delivery system, advocates of this approach talked about achieving greater efficiency in the delivery of education services—consolidating school districts, monitoring school districts more closely, setting state standards, and holding schools accountable for attaining the objectives. (Sederburg, 1994, p. 3 of 8; emphasis added)

The first two approaches, choice/competition and systems, were widely known and publicly debated.[5] The management approach, however, was a term unknown to the public at large, and the term had not circulated in public debate about education. The position of "management approach" was initiated by the survey report itself! The report not only initiated the use of the term in Michigan educational discourse, it also generated and codified the management approach as a significant factor in political alliances for educational governance:

> The Michigan Education Poll finds the public divided about what approach to use to improve education. With the primary responsibility for funding schools shifted to the state, the debate over improvement strategies is likely to intensify. It remains to be seen if advocates of improving the system form a majority coalition with supporters of greater management or if advocates of choice coalesce with supporters of the management approach. (Sederburg, 1994, p. 5)

The Michigan Partnership deployed Public Sector Consultants as a way to mobilize a kind of cooperative participation that resonated with and bolstered the appeal of the Third Way. With the deployment of the public opinion polls and the promotion of the management approach, the Partnership discursively reiterated a particular form of democratic participation.

The rhetoric of the Third Way is seductive because it appears to bypass not only the compromises of centralized state leadership, but also the selfishness of decentralized liberal individualism. However, Third Way thinking embodies its own disciplinary mechanisms. Nikolas Rose's critique of Third Way thinking focuses on the construction of the citizen as a moral subject, an examination of what he calls "ethopolitics" (Rose,

2000). Rose explicates a "double movement" in the shifts in governance patterns from the state and civil society to a notion of partnership:

> Organization and other actors that were once enmeshed in the complex and bureaucratic lines of force of the social state are to be set free to find their own destiny. Yet, at the same time, they are to be made responsible for that destiny, and for the destiny of society as a whole, in new ways. Politics is to be returned to society itself, but no longer in a social form: in the form of individual morality, organizational responsibility and ethical community. (Rose, 1999, pp. 174–175)

This shift in governance constructs forms of citizenship that depart from more traditional notions of social and democratic participation. The discourse of partnership then, becomes a vehicle of self-governance by which citizens can envision their participation in ways that appear independent and nonpartisan, while at the same time embodying accountability to particular visions of a managed educational system. Third Way thinking paves the way for educational governance to be instantiated as a new site for regulating citizenship, while appearing to operate outside the structures of government. Membership and participation in any given partnership enact the simultaneous processes of inclusion and exclusion (Popkewitz and Bloch, 2001; Popkewitz, 1998). The public opinion poll and the management approach appear to occupy an inclusive middle ground, so the partisanship, mechanisms of exclusion, censorship, and normalization are obscured.

Funding: Equity and Control

Before 1993, Michigan public schools were funded as they are in most other U.S. states: district-by-district on the basis of local property taxes. When the district-based property tax funding pattern is in place, schools in wealthy districts have more money than schools in poor districts. In an effort to establish a more equitable distribution of public money, on July 22, 1993, Michigan residents voted in favor of Proposal A, which rejected property taxes as the basis of support for public education, and shifted the distribution of public school money away from local districts and into the governor's office. Interestingly, at the time of the vote, an alternative funding mechanism to support public education had not yet been determined.

A selection from the 1994 opinion survey illustrates how complex perspectives on school funding were portrayed in Michigan in the mid-1990s:

> Passage of Proposal A ensures that funding will shift from being primarily a local responsibility to being a state responsibility. Consequently, state

policymakers may be tempted to establish greater control over the educational system.

The 1993–94 Michigan Education Poll asked the public whether they believe that each of four activities [auditing financial records; establishing performance standards; establishing the types of educational programs offered by schools; and providing parents with additional choices in the schools available to their children] is a legitimate state function or if the activity should be left to the local school district.... Slightly more than one in four respondents (27 percent) believe the state should not play a role in any of the activities. In contrast, only 11 percent believe the state should do all four activities. The Michigan public remains committed to local control in education. (Sederburg, 1994, p. 2)

This last paragraph represents an interesting and not precisely wrong way to aggregate the data. However, Exhibit 7 of the report on which this paragraph is based indicates that the majority of those polled favored state over local control in two of the four activities—auditing financial records and establishing performance standards (Sederburg, 1994, p. 7). Rhetorically, the survey analysis repeats the assertion that "the Michigan public remains committed to local control in education" even though the data in the same report are less conclusive. Further confounding the issue is that the passage of Proposal A shifted control of school funding precisely away from local districts and directly into the hands of the state. The preceding excerpt is illustrative of the way the discourse about funding was recast to eliminate the traditional rhetorical opposition between centralization and decentralization.

The shift in funding mechanisms to the state, together with the rhetorical spin that made it sound like a move toward "local control" can be seen in conjunction with the technologies of public opinion polling, and promotion of a management approach to governance. These components of the Michigan Partnership are reflective of how partnership discourse redefined civic participation and recast the language of debate.

Traditionally in the United States, the discourse of citizenship has been seen as a choice between individualist direct-democracy participation and collectivist social welfare responsibility. However, the rhetoric of the public opinion poll, including its emphasis on local control and popular support for a management approach, contributed to changing the language in which it is possible to talk about civic participation. Partnership discourse constructed state-based funding as local control, and cast opinion-polling expertise in the role of the "voice of the people." In this way, the discourse of the Michigan Partnership instantiates the fabrication of historically specific forms of citizenship.

School Types: From Professional Development Schools to Charter Schools
Initially, the central focus of the Michigan Partnership for New Education was the establishment of Professional Development Schools (PDSs), which would be the site of university/school cooperation. However, between 1993 and 1994, the momentum for educational reform shifted dramatically away from PDSs and toward charter schools. Given the enthusiasm and resources available to the Partnership, it may surprise some that support for PDSs died out so quickly. This shift in direction is especially noteworthy given the fact that the 1994 Michigan Education Poll reported explicitly that charter schools were not highly regarded by the Michigan public. In a section entitled "Public Perception of the Effect of School Reform," the report concluded, "school accreditation and charter schools were not seen as a very effective method for improving schools" (Sederburg, 1994, p. 4). Next, I would like to highlight three sorts of factors that occurred in the Michigan Partnership between 1993 and 1995: a series of legislative actions and court cases, changes in key personnel, and the introduction of for-profit educational corporations into some school districts. I point to these three factors as a way of analyzing a particular redirection in educational reform.

In the first of these three factors, a "very curious" (Furst, 1996) series of legislative actions and court cases contributed to the changing tide of reform that allowed the terms of the debate to shift away from the traditional opposition between centralization and decentralization. Through a series of legislative moves, Michigan educational partnerships forged particular sets of relations to govern public schools. On July 22, 1993, the Michigan legislature voted for a tax cut that removed from school operating budgets, more than $6 billion, from the property tax base[6] (Furst, 1996; Geltner, 1994). While the Michigan legislature was working on an alternative school funding plan to replace the property tax base, the governor "announced that he would not sign any funding bill unless major educational reform were being considered" (Furst, 1996, p. 233).

The governor's insistence on major educational reform was motivated partly by the public outcry in response to *A Nation at Risk*, in which U.S. schools were portrayed as being in a state of crisis. Wanting to be seen as rising to this challenge, but not wanting to give the impression of big government and state control, the Michigan legislature quickly enacted PA (Public Act) 284, which provided for charter schools, which the legislature called "public school academies." The first school to apply for a charter was Noah Webster Academy, which was comprised of several hundred home-schools around Ionia, Michigan.[7]

Charter schools were not unanimously popular, however. Some constituents in Michigan saw charter schools as a thinly disguised step in the direction of school vouchers and public funding of private schools. Objecting to the legislative establishment of charter schools (PA 362), the teachers' union (Michigan Education Association), together with others, filed suit claiming that charter schools violated the state constitution.[8] Before the presiding judge, William Collette, the plaintiffs raised three points on which charter schools were challenged as being unconstitutional.[9] After PA 362 was found unconstitutional in the county court, the legislature responded by enacting PA 416 in 1994, which provided new restrictions that would make it more amenable to the court rulings. Most notably PA 416 stipulated that the State Board of Education would be the governing body to which charter schools would have to answer.[10]

This flurry of legislation and an unfavorable court ruling is indicative of the unrest and transitional nature of 1990s school reform agendas. Support for charter schools was articulated in Third Way rhetoric that seemed to satisfy desires for both populist and expert authority. For example, charter school policy promised to "expand the number of authorizing bodies" and "open the doors for innovative teaching." At the same time, charter schools were governed according to state-centered funding policy and state School Board regulation.

The shift in school types from PDSs to charter schools was constituted by legislation, court cases, public polling reports, and public concern generated by *A Nation at Risk*. In addition, personnel changes in key Partnership positions were another significant factor in the reform directions that allowed charter schools to replace PDSs. In 1989, James Blanchard was governor of Michigan, A. Alfred Taubman (a shopping-mall developer) was the major financial backer, and Judith Taack Lanier (dean of the college of education at Michigan State) was the executive director. A year later, Engler replaced James Blanchard as governor of Michigan. Three years after that, in 1993, Taubman appointed Harrison Blackmond (a California lawyer and director of Human Services at the Taubman Company) to be executive vice-president and CEO of the Partnership. But then, in September 1994, Taubman hired William Coats (former professor of education and school superintendent) to be president and CEO of the Partnership replacing both Blackmond and Lanier. The sequence of personnel changes from the governor to the Partnership directorate synchronized with the shifting discourse of reform in favor of charter schools.

The new governor was instrumental in the shift from PDSs to charter schools. In his 1990 bid for the governorship, Engler had campaigned in favor of a smaller-government, market-based strategy for educational reform. While in office, his administration continued to support the Michigan Partnership by budgeting $5.3 million in state monies for 1994, however, Engler pushed heavily for more participation by the private sector, and a more choice/competition approach to public education. The state government was supported by an atmosphere conducive to private educational initiatives, and some local districts also took action. For example, on January 10, 1994, the Board of Education of Pinkney, Michigan privatized the entire K-12 district under the auspices of a for-profit corporation, Education Alternatives, Inc. (Schmidt, 1994).

This district-based privatization move was another indication of how partnership language allowed the blurring of the distinction between centralization and decentralization. On the one hand, the district appeared to have acted independently of the State Department of Education. On the other hand, the district found plenty of supportive rhetoric to be able to justify hiring a corporation that assumed responsibility for school management, organization, policy, and curriculum. Going with the flow of the Third Way tide, school reform operations had now been transferred first from local school districts to the state, and then from the state to a private corporation. This shift was not debated according to conventional political positions of centralization and decentralization. Rather, in accordance with the language of the polls, the legislative proposals, and the court cases, it became thinkable—even reasonable—to transfer decision-making to a Minneapolis-based corporate entity as remote from populist democratic input as the state allegedly had been.

The move of local school districts to contract professional corporations for purposes of school reform is another instance of the reconstruction of possibilities for participation away from traditional centralist or decentralist models, and in accordance with partnership rhetoric. The discourse of the Third Way helped to normalize the possibility of privatization by making traditional forms of participation seem partisan and obsolete. The rhetorical possibilities afforded by Third Way thinking provide the basis by which privatization can be understood to be within the public sphere, and corporate expertise seems to be the same as local control.

In a move that was later condemned as conservative, Michigan State University refused to support the Michigan Partnership's shift of emphasis away from PDSs and in favor of charter schools. Voicing their objections to the charter school movement, representatives from MSU denounced the Partnership's effort to establish a statewide network of charter schools. While several state universities became authorizing agencies for charter schools, MSU refused to accept charter applications. Maintaining her original stance in favor of professional development schools, Judith Lanier, a key leader and spokesperson from MSU, left the Michigan Partnership in 1994.

The Michigan Partnership is a case in which discursive moves constituted an overhaul of the traditional opposition between centralized and decentralized systems of policy and regulation. Three realms of analysis—governing bodies, funding patterns, and school types—illustrate how centralization and decentralization practices can be interwoven to construct a discourse that allows for social fabrications in which the traditional meanings of "local control" and "by the people" do not mean the same things as they used to. In the case of the Michigan Partnership, the particular network of relations was constituted by the activism of the Public Sector Consultants, the curious series of legislation and court cases, the shifts in administrative priorities, and the historically specific set of power relations among the state, the university, the Partnership, the teachers' union, and the polling entities. In the next section, the analysis of partnership discourse focuses on shifting relations within two sets of traditional oppositions: between democracy and efficiency, and between standards and reform.

Redoubling Tensions: Accountability for Processes and Products

In conventional terms, school accountability has been measured according to two very different sets of criteria: either on the basis of adherence to a standardized process (the *input* model), or on the basis of attainment of predetermined objectives (the *output* model). In efforts to maintain fair and just democratic procedures, universities have tended to favor process-based accountability (Mintrom, 2001), while corporate entities tend to favor more efficient output models of accountability. In 1987, the Michigan State Board of Education set forth the *Model Core Curriculum Student Outcomes*, an instance of the shift in accountability from process to product. Outcomes-based accountability had been gaining prominence in educational sectors for several decades. During the Partnership's early years, a culture of process-based accountability was formally displaced by outcomes-based education and bottom-line management approaches. Quoting the *Model*, Geltner (1994) writes, "the philosophy and rationale of the model reflected the growing national focus on educational outcomes rather than inputs, shifting the emphasis from what was *taught* to students to what was *learned* by them, creating 'a new accountability for education based on results, not intentions' " (p. 403; emphases in original).

The shifts in the bases for accountability inscribe broader historical shifts from process-based to product-based governance. They also resonate with the long-standing discourse that posits a debate between efficiency and democracy as conflicting purposes of school reform (see e.g., Tyack, 1974; Kliebard, 1986; Labaree, 1997; Tyack and Cuban, 1995). The process-based pattern of organization is compatible with Fordist assembly lines that were organized as a series of ordered steps or procedures. In the case of process-based governance, the sequence of steps is stipulated, so the administrator is not "free" to improvise a method or approach; the steps on the assembly line are fixed. In process-based accountability, the methodological steps are controlled, but the outcome is not explicitly stated, and the outcome is not the basis for accountability measures.

In contrast, when educational reform is defined in terms of objectives, the converse is the case. Product-based (output-, objectives-, or outcome-based) educational accountability leaves the procedure relatively unspecified, but the eventual outcome has been stated in advance. In an output-based accountability scheme, school reforms are evaluated on the degree to which they attain the explicitly stated predetermined objectives. Product-based reforms in education are commensurate with bottom-line management, target-quotas, and outcome-based mechanisms in related fields. Outcome-based accountability in social and political sectors is often (vulgarly) called "pragmatism" because the procedures are subordinated to—and determined by—the goals. Reflecting corporate preferences, the Partnership embraced the discourse of outcomes-based accountability. In contrast, Michigan State University continued to adhere to a program of professional development schools and process-based accountability. Failure to reconcile these two approaches led eventually to MSU's departure from the Partnership.

When the Michigan Partnership got wrapped up in outcomes-based reform, the nature of educational accountability was also redefined using a Third Way kind of rhetoric. In this rhetoric of accountability, the state would stipulate educational outcomes to be measured by standardized tests. However, state control was made to seem innocuous because local school districts were to be left free to employ any methods or processes they chose in order to meet those objectives. Procedures were left unspecified, and that made it seem like local school districts could act autonomously. Again, in this accountability arrangement, the traditional opposition between centralized state control and decentralized local control has been undone. In its place is a system of accountability in which local control is understood to mean control of procedures, while the state has jurisdiction over the stipulation of outcomes.

One major impact of this redefinition of accountability is to eliminate the possibility of unanticipated outcomes. That is, in a process-based accountability scheme, there is always the real and theoretical possibility of surprising results. Although not always desirable, surprising results can be the source of genuine innovation. In contrast, however, the outcomes-based accountability scheme forecloses the possibility of new or surprising results. Even if there is local control of procedures, an outcomes-based scheme, accountability can be measured only on the basis of predetermined objectives. Therefore, at some level, there is no longer any possibility for innovation, or ironically, for reform.[11]

School Improvement Plans
In the Michigan Partnership, a system of accountability was instituted that reconstructed the notion of accountability in a way that conformed neither to process-based nor to product-based accountability in the traditional senses. Part of this reconstruction of accountability was generated by the Partnership in response to a state Revised School Code called the "School Improvement Plans" or SIPs. SIPs were mandated by PA25 in 1990 by the Michigan legislature. The state code for SIPs provided a framework for accountability that conformed to neither the process-based nor the product-based systems of accountability. Rather, this accountability framework combined aspects of both traditional approaches into a different sort of discursive framework. The text of PA 25 reads as follows:

> The school improvement plans shall include, but are not limited to, a mission statement, goals based on student academic objectives for all students, curriculum alignment corresponding with those goals, evaluation processes, staff development, development and utilization of community resources and volunteers, the role of adult and community education, libraries and community colleges in the learning community, and building level decision making. School board members, school building administrators, teachers and other school employees, pupils, parents of pupils attending that school, and other residents of the school district shall be invited and allowed to voluntarily participate in the development, review, and evaluation of the district's school improvement plans. (Michigan Legislature, 1990)

In this document we can see elements of process-based accountability in the stipulation that representative from a variety of social sectors be included in the decision-making and review processes. We can also see elements of product-based accountability in the emphasis on attainment of established goals. The language of this policy document both reflected and contributed to a discursive framework that made it possible to talk about accountability in a way that did not resort to conventional oppositions between input and output models.

In the Michigan Partnership's work with schools, SIPs became the central focus of reform (Peters, 2002). By 1993, support for professional development schools had begun to wane, funding was down, and the new state administration supported charter schools and did not support PDSs. The SIPs, then became the focal point for Partnership efforts in schools because School Improvement Plans had become the formal mechanism for school reform accountability. In addition to the language of the Revised School Code Act, SIPs became technologies of accountability that folded output expectations together with input initiatives to govern and normalize possibilities for reform.

One significant aspect of the 1990 Revised School Code was that SIPs were required every year:

> if the board of a school district wants all of the schools of the school district to be accredited under section 1280, the board shall adopt and implement and, not later than September 1 each year, shall make available to the department a copy of a 3- to 5-year school improvement plan and continuing school improvement process for each school within the school district. (Michigan Legislature, Revised School Code, Act 451)

The School Improvement Plan became the documentation on which the state determined accountability. The plan became the target; the process became the product; the input became the output.

Accountability for Reform and Reform of Accountability
The Michigan Partnership is a case that helps illuminate historical trends in educational partnerships. In this section I address general issues of governing principles that circulate through educational reforms.

One effect of School Improvement Plan policy was to shift the attention, energy, and resources of Michigan Partnership endeavors into the continual production of and documentation for SIPs. The result was the construction of intense monitoring and surveillance of the school reform efforts. The requirements of the School Improvement Plan generated a culture of continual assessment in which school improvement became coterminous with accountability for an administrative plan for improvement. SIPs can be seen as an instance of what Deleuze (1992) has called "societies of control" (in contrast to "societies of discipline"):

> The administrations in charge never cease announcing supposedly necessary reforms: to reform schools, to reform industries, hospitals, the armed forces, prisons. But everyone knows that these institutions are finished, whatever the length of their expiration periods. It's only a matter of administering their last rites and of keeping people employed until the installation of the new forces knocking at the door. These are the *societies of control*, which are in the process of replacing the disciplinary societies. (Deleuze, 1992, p. 4; emphasis in original)

The contrast between the idea of a *disciplinary society* and *control society* inscribes many of the aspects that characterize the discourse of the Michigan Partnership. Deleuze suggests that new or emerging patterns of power relations are sufficiently distinct from the relations of modernity, that a society of discipline no longer pertains to all aspects of society, and that the emerging power relations constitute societies of control. I understand Deleuze's control society as different from a disciplinary society in three respects: (1) both discipline and control societies are characterized by the self-monitoring gaze; but in a control society the monitoring is more frequent and continuous than in a disciplinary society; (2) standards in a disciplinary society tend to be fairly centralized and long-lasting; however, standards in a control society are more heterogeneous and quickly changing; (3) a disciplinary society afforded the promise of closure or completion of a project; however, a control society offers no possibility of closure or completion.

The first salient aspect of the disciplinary society that is now different in the control society is in the nature and rhythm of its regulatory mechanisms. In a disciplinary society, the outcome, or product may be evaluated only once, perhaps by a final exam or quality control unit at the end or completion of a session. At the end of the term or assembly line, students or factory products are inspected, tested, and evaluated. The members of a disciplinary society are self-disciplined and productive members of society. The results of a Five-Year School Reform plan would be evaluated after five years.

According to Deleuze, monitoring in a control society is more frequent than in a disciplinary society. A control society is characterized by *continuous monitoring*: "Indeed, just as the corporation replaces the factory, *perpetual training* tends to replace the school, and *continuous control* to replace the examination" (Deleuze, 1992, p. 5, first emphasis in original; second emphasis added). In schools, there is evidence of a shift from grading on the basis of a final exam to grading on the basis of many more frequent tests throughout the semester. Smaller, weekly papers are replacing the "one big" research paper required in previous decades. SIPs require a plan to be written every year, even though every report must stipulate plans for three to five years in the future. The shift in frequency of monitoring is evident in fields other than education, including the move in economics when a fixed gold standard was replaced by floating rates of exchange; in criminology electronic tracking devices are locked onto "prisoners" rather than having prisoners locked up in a (fixed place) prison; and in business, marketing in the form of continuous multimedia advertising is replacing brand-name loyalty and market niches. This notion of continuous monitoring characterizes the Partnership's enactment of School Improvement Plans, as well as the increased frequency of public opinion polling and reporting.

The second aspect is in the heterogeneity of standards in a control society. Standards in a disciplinary society could be regarded as relatively centralized or uniform. In contrast, a control society is one in which "standards and demands can come from anywhere at any time, in any form" (Ball, 1999). For example, a school curriculum is no longer accountable only to school-board criteria of education. School curricula are now also answerable to partnership constituents that include businesses, churches, parents' groups, social service providers, psychiatrists, and police forces. In order to manage a classroom, teachers must be familiar with a wide range of experts in order to make appropriate referrals for children to social services, parent representatives, community liaisons, and legal services. Education must be understood to serve a multicultural, multilingual, and culturally fragmented constituency. In some places, other members of the school community include the McDonalds or Taco Bell franchises that operate in the school lunchrooms (Kaplan, 1996). The Michigan Partnership is clearly a case in which school reform was held accountable according to heterogeneous standards—like public and private, or centralized and decentralized—even as those standards were woven together in a discourse that made them seem compatible.

According to Deleuze, the final contrast between the discipline society and the control society is in the possibility for completion. In a disciplinary society, one could graduate or be promoted to another rank. However, in a control society, he notes, completion is not an option:

> In the disciplinary societies, one was always starting again . . ., while in the societies of control one is never finished with anything—the corporation,

the educational system, the armed services being metastable states coexisting in one and the same modulation, like a universal system of deformation. (Deleuze, 1992, p. 5)

The notion of "never finished" is also inscribed in recent partnership programs of "whole-life" and "lifespan" development, as well as in "life-long" learning and "continuing education" programs in schooling. One never graduates; one never completes an education. These technologies of the control society—continuous monitoring, heterogeneous standards, and never finished—are deployed in the Michigan Partnerships for the ostensible purpose of raising standardized test scores and improving the educational outcome for Michigan's children.

Conclusion

Certainly all sectors of society are invested in the welfare of the educational system. So it is not surprising that a partnership of state, community, business, and university interests sounds like a promising innovation to effect much needed educational reform. Perhaps the case of the Michigan Partnership for New Education is just another example of "easier said than done," or perhaps it is a case of "business as usual." In any case, this chapter offered an analysis of the mechanisms through which the Michigan Partnership worked through a time of significant reform. The analysis provided a critical perspective of how a partnership can engage in power relations, and how those power relations work to recast the terms in which it is possible to talk about reform and participation. Most importantly, as entities like the Partnership foster relationships among various social sectors, they open up new venues for civic participation—public opinion polls, Third Way thinking, blurring of centralization and decentralization, process-as-product accountability, and corporate expertise as the voice-of-the-people. At the same time, traditional venues of participation and debate are dissolved—for example, the rhetorical distinction between direct-democracy individualism and representative social collectivism. While the traditional terminologies of "local control" and "popular support" may still permeate the discourse of reform, the meanings of those terms changes when entities like the Partnership embody new sets of relations among participating voices.

By 1995, the educational reform agenda of the MPNE had altered radically from its original goals. In 1996, the leadership of the Partnership made a move to reflect that change. The MPNE was dissolved, and in its place was born the Leona Group, a fully privatized corporation that owns and manages schools they call "public charter schools." According to their website, the Leona Group is based in both Michigan and Arizona:

> The Leona Group, L.L.C., was formed in 1996 in Michigan by William Coats, Ph.D., a national leader in public education reform. With a vision of creating "a new kind of public school," The Leona Group's mission is to promote choice and competition in public education for all students, including those with special needs, to help them pursue their individual academic goals. (The Leona Group, L.L.C., 2002)

William Coats, former chair of the Partnership founded the Leona Group, but the latter corporation does not make any pretense of partnership or cooperation with universities, states, or local school districts. Its rhetoric is no longer one of broad-based coalition and innovation. Rather, the Leona Group is organized as a for-profit corporation, and their rhetoric of school improvement is fully market-based, "to promote choice and competition in public education."

It is unlikely that the founders of the Michigan Partnership could have anticipated the series of legislative events, personnel changes, and changing tide of public opinion that eventually transformed the notion of Partnership from a vision of civic participation and into a private corporate endeavor. The eventual transformation was not instigated from a single source, action, or person. Rather, it entailed interactions on multiple fronts over several years. The Michigan legislature, the courts, the teacher's unions, the university, the Holmes Group, local school districts, corporate sponsors, *A Nation at Risk*, and the voters of Michigan all participated in a complex series of interrelations, initiatives, and responses that shaped and reshaped the Partnership agenda of educational reform. It remains to be seen whether corporate entities like the Leona Group continue to function as emblems of "choice and competition" for educating children, or whether they will be regarded as representatives of the private sector in a broader debate with other social sectors for educational improvement.

Notes

I would like to thank Joyce Grant, Susan Peters, and Doug Campbell for their support and for sharing information about their first-hand experiences in the Michigan Partnership. All misrepresentations of that information are my own.

1. The Holmes Group produced three nationally influential documents: *Tomorrow's Teachers* (1986), *Tomorrow's Schools* (1990), *and Tomorrows Schools*

of Education (1995). The Holmes Group is now known as the Holmes Partnership and is based in Waco, Texas and Washington, DC.
2. For example, Judith Taack Lanier was president of MPNE, president of the Holmes Group, and dean of the College of Education at MSU (on leave during 1990) at the same time.
3. Perhaps the most notorious public figure is the Partnership's organizer and supporter, A. Alfred Taubman, who was also chairman and director of Sotheby's prestigious auction house.
4. "Two approaches to improving education received much attention during the debate over education reform. One approach (referred to in this report as the *choice/competition* approach) was emphasized by Governor Engler in his September 1993 speech before the Michigan legislature. The governor stressed the value of offering parents increased choice in the schools their children can attend . . . The second approach that policymakers discussed was to improve education by strengthening the current system (that is, the *systems approach*). This approach emphasized the need for the state to supply schools with adequate resources, adopt school improvement plans written by local educators, improve teacher education, and empower education professionals . . ."
5. According to the survey, the choice/competition approach was favored by residents of central and southeast Michigan. The systems approach was favored by education professionals and people with higher levels of education. Actually, preferences for these approaches include many subgroups. I indicate here only salient representative categories. The report includes the complete polling data.
6. The bill was signed into law on August 19 by Governor John Engler.
7. Most charters were submitted to central Michigan University, however, the Noah Webster charter was granted by the local public school district.
8. Council of Organization and Others of Education About Parochiaid Inc. v John Engler, No. 94-78461-AW. Ingham County Circuit Court, November 1, 1994.
9. "Are academy schools, as provided for in 1993 PA 362, 'public schools' for which state funding is available?" The judge used court precedents concerning Indian schools to rule that academies were not the same as public schools. "The public school must (1) be under the exclusive control of the state of Michigan and (2) must be open for enrollment to all children within the district where it is located" (Furst, 1996, p. 239). An academy, however, would be run by a private board as are other nonprofit organizations, the State Board of Education was not responsible for oversight, and academies were under no restrictions about teaching religion. "Does 1993 PA 362 violate the constitution by declaring academy school to be 'school districts' under some provisions of the constitution, but not others?" Plaintiffs argued that charters granted private corporations the ability to levy taxes, which is unconstitutional. Judge Collette refused to rule on this aspect citing lack of precedent. "Does 1993 PA 362 violate article 8, § 3, of the Michigan constitution of 1963 in that it divests the State Board of Education of its duty to lead and supervise public education in Michigan?" The original charter

school legislation, PA 284, provided that the State Board of Education would be the oversight body, however, PA 362, which superceded PA 284, removed that provision. Supporters of charter schools argued that PA 362 did not prevent the State Board of Education from exercising oversight, it just didn't require that oversight. Charter school oversight lay with the authorizing body, so Judge Collette ruled that provision unconstitutional.

10. Interestingly, on July 30, 1997, the Michigan Supreme Court reversed the previous Ingham County Circuit Court decision. The Supreme Court asserted that "public school academies are under the ultimate and immediate control of the state and its agents" (State of Michigan, 1996–1997:6), therefore "they are necessarily subject to leadership and general supervision of the State Board of Education to the same extent as are all other public schools" (p. 7).

11. See Mintrom (2000), for an extensive discussion of how charter schools have failed to generate innovation in Michigan school reform.

References

Ball, S. (1991). "Policy Sociology: Critical and Postmodern Perspectives on Education Policy." Havens Center Lecture, Madison, WI, February 24, 1999.

Bradley, A. (December 12, 1990). "M.S.U. Education School is on a Mission: 'Teaching for Understanding.' " *Education Week*, from http://www.edweek.org.

Deleuze, G. (Winter 1992). "Postscript on the Societies of Control." *October*, 59, 3–7.

Etzioni, A. (2000). *The Third Way to a Good Society*. London: Demos.

Foucault, M. (1979). *Discipline and Punish: The Birth of the Prison*. New York: Vintage.

Furst, L.G. (January 25, 1996). "The Short but Very Curious Legal History of Michigan's Charter Schools." *Education Law Reporter*, 105, 233–245.

Geltner, B.B. (1994). "Juggling the Reinvention of Michigan's Public Schools." *International Journal of Educational Reform*, 3(4), 401–413.

Giddens, A. (1998). *The Third Way*. Oxford, UK: Blackwell.

Kaplan, G.R. (1996). "Profits R Us: Notes on the Commercialization of America's Schools." *Phi Delta Kappan*, 7(3), K1–K12.

Kliebard, H.M. (1986). *Struggle for the American Curriculum*. London: Routledge & Kegan Paul.

Labaree, D.F. (Spring 1997). "Public Goods, Private Goods: The American Struggle over Educational Goals." *American Educational Research Journal*, 34(1), 39–81.

Michigan Legislature. (1990). Revised School Code. Public Act 25 (380.1277).

Michigan Partnership for New Education. (September 1993). *1993–1994 Plan*. Vol. 1. Submitted to the Michigan Department of Education.

Mintrom, M. (2000). *Leveraging Local Innovation: The Case of Michigan's Charter Schools*. Michigan State University: Author.

Mintrom, M. (November 2001). "Educational Governance and Democratic Practice." *Educational Policy*, 15(5), 615–643.

Peters, S. (2002). "Inclusive Education in Accelerated and Professional Development Schools: A Case-Based Study of Two School Reform Efforts in the USA." *International Journal of Inclusive Education*, 6(4), 287–308.

Popkewitz, T.S. (1998). *Struggling for the Soul: The Politics of Schooling and the Construction of the Teacher.* New York: Teachers College Press.

Popkewitz, T.S. and Bloch, M. (2001). "Administering Freedom: A History of the Present: Rescuing the Parent to Rescue the Child for Society." In Hultqvuist, K. and Dahlberg, G. (Eds.), *Governing the Child in the New Millennium* (pp. 85–118). New York: RoutledgeFalmer.

Rose, N. (1999). *Powers of Freedom: Reframing Political Thought.* Cambridge, UK: Cambridge University Press.

Rose, N. (June/July 2000). "Community, Citizenship, and the Third Way." *The American Behavioral Scientist*, 43(9), 1395–1411.

Schmidt, P. (January 19, 1994). "E.A.I Reaches Tentative Accord to Run Michigan Schools." *Education Week*, from, http://www.edweek.org/ew/ew_printstory.cfm?slug=17eai.h13.

Sederburg, W. (June 3, 1994). *The 1993–93[sic] Michigan Education Poll: Focus on Reform.* Lansing, MI: Public Sector Consultants, Public Opinion Research Institute, from, http://www.publicsectorconsultants.com/PSR/ Adv/1994/ 060394.cfm.

The Leona Group. (2002). *Background*, from, http://www.leonagroup.com/ boct.html.

Tyack, D. (1974). *The one Best System: A History of American Urban Education.* Cambridge, MA: Harvard University Press.

Tyack, D. and Cuban, L. (1995). *Tinkering Toward Utopia: A Century of Public School Reform.* Cambridge, MA: Harvard University Press.

Part IV
Governing as a Problem of Inclusion and Exclusion

Chapter 8
Partnerships and Parents: Issues of Sex and Gender in Policy and Practice

Miriam E. David

Introduction

In this chapter I explore the notion of partnerships as it has emerged as part of global public policy discourses under anglophone regimes of neoliberalism in the late twentieth century. Specifically I want to interrogate its origins and uses in Britain whilst also being sensitive to how the notion has emerged and become embedded and embodied in such practices in the United States. Equally I want also to explore how the notion has been exported from the United States to parts of Europe as a form of neoliberal colonialism and has incorporated new discourses of policy and practice in European networks, especially in relation to educational discourses.

As far as we can tell, the notion of partnerships as part of global official policy discourses emerged at the same time that policy-makers and social science researchers were beginning to question transformations in family life away from traditional forms of marital relations and the nuclear family. These took diverse forms of adult relations toward forms of consensual unions, cohabitation partnerships and similarly diverse forms of parent relations with children. The idea of partnerships can be associated with these "personal" transformations in respect of its use in public discourses. However, its adoption and entrenchment has a parallel and separate embodiment in legal and business discourse, and in some of the more innovative connotations about liberal forms of business practices as applied to education (Whitty, 2002). Thus the notion derives from several separate sources and discourses, most particularly notions of the introduction of market forces and the marketization of public services, most especially education.

This coupling of the public adoption of the notion of partnerships with personal, private, and familial changes entails wide and deep cultural transformations. Thus the contradictory processes of social change and sexual transformations are both embodied and embedded in these deeper discursive shifts. I want to explore two facets of these broader shifts in the discursive lexicon of New Labour in Britain by comparison with the United States. It has adopted and assimilated this cultural and sexual language in terms of partnerships between families and education, schools especially, and within the cultural and pedagogic practices of schools, through their curricula about sexuality, sex, and relationships.

The specific but broader context for these shifts relates to how New Labour adopted and adapted the social and sexual agendas of neoliberalism under conservative regimes. Indeed, in a curious mix of policy borrowings (Silver, 1994) from diverse regimes, New Labour adapted conservative policies from "Majorism" and "Thatcherism" and at the same time tried to borrow from and adopt more liberal agendas from the New Democrats in the United States. In both Britain and the United States, there had been moves away from politically liberal regimes of the 1960s and 1970s associated with either social democracy in Britain or "social liberalism" in the United States toward economic liberalism or conservatism in the 1980s and 1990s—Thatcherism in Britain and Reaganism in the United States. However, neither the New Democrats in the United States nor New Labour in Britain contemplated a straightforward return to old political values and ideologies of social liberalism and notions of social and economic inequality. In both respects their shifts have also entailed discursive shifts toward notions of social exclusion and its corollary social inclusion rather than traditional concepts (Levitas, 1998).

Many critical thinkers and intellectuals have noted the complexity of the discursive shifts under regimes of neoliberalism, pointing to the cultural turn and political revisions in conceptions of politics and power. I will try to follow Nancy Fraser (1997, 2000) who has attempted to fashion a critical social and feminist theory as a way of providing a general political–intellectual orientation to what she calls "the post-socialist condition." She argues that "a critical social theory of recognition is a specific project to be undertaken" (2000, p. 204) to develop a socialist perspective as a space for thinking that does not "get closed off prematurely by neoliberal ideology" (2000, p. 205). Thus she attempts to combine a revised socialist perspective with feminist and post-structuralist theories as a way to critique politics and policies within regimes of neoliberalism. This essay is an attempt to analyze critically educational discourses about partnerships in Britain by comparison with the United States and their complexity in relation to gender and sex in this specific "post-socialist condition" and context.

The Political and Personal Turn

The intellectual justification for the political challenges to both traditional conservative and socialist ideologies and practices emerged slowly in Britain through what was initially known as New Labour's modernization project into its elevation into the politics of the Third Way in Britain. The original political ideology of New Labour was initially formulated through the Institute for Public Policy Research (IPPR) a left-leaning public policy think tank in Britain. The Labour party and its late leader, John Smith, set up a quasi-official organization whilst an opposition party to government, namely the Commission on Social Justice (1994) based within the IPPR that renegotiated New Labour's political principles and practices. It was slowly sketched into a political project of modernization bringing together notions of market forces within public services when Blair took over as leader of the Labour party (Blair, 1998).

The notion of Third Way was given sustenance and legitimacy through the writings and publications of Britain's most famous sociologist, Professor Anthony Giddens and a member of IPPR. He had become director of the London School of Economics only six months before New Labour was elected to office, having been the first Professor of Sociology at King's College, Cambridge University. He turned these embryonic ideas into an intellectually respectable project through being appointed an unofficial advisor to Prime Minister Tony Blair and publishing them as a book. The flyleaf of his book (1998) stated: "Frequently referred to in the UK as Tony Blair's guru, Giddens has made a strong impact on the evolution of New Labour" (1998, 2000). He argued for the development of a new politics that neither adhered to traditional conservative notions of economic liberalism and freedom from political intervention, nor traditional socialist notions of social and economic equality. His Third Way sought to weave a path through these ideological and political concepts and political practices to a combination of the inevitability of economic markets, regulated by social constraints and political interventions. His emphasis, however, was on what he called "the renewal of social democracy," that is a return to traditional social democratic ideals of equality, rather than a full commitment to the marketization of all public services.

He also paid attention to the social and cultural transformations that had taken place since the fashioning of the traditional political ideologies of left and right (Giddens, 1994). For instance, in a commissioned series of prestigious public lectures broadcast on the radio as the now famous Reith lectures he also developed notions about the scale and extent of global social transformations. He entitled the public lectures *Runaway World* (1999) to demonstrate these varied and various processes of economic and social change. The title was borrowed from a series of Reith lectures presented 30 years earlier by Sir Edmund Leach, a professor of social anthropology at Kings College, Cambridge. Leach had also called into question the extent of social change and its effects on private family lives, giving consideration to how the family's "tawdry secrets" remained essentially private rather than public. Giddens, however, asserted that there had been massive social and familial transformations over the subsequent 30 years, which crucially included transformations in family, personal life, and intimacy. These meant that such changes had become critical public rather than private and intimate matters, and subject to social and public scrutiny.

Giddens's ideas about love, intimacy, and marriage were developed from both his previous work (1992) and the various feminist and politically conservative commentaries on family life changes (David, 2003). Giddens emphasized the transformations in social and sexual relations, paying most attention to shifts in personal lives and forms of love and intimacy, rather than the more structural shifts in patterns of marriage and divorce in late modern societies, such as Britain and the United States. Nevertheless, he did acknowledge the transformations toward new forms of *adult partnerships* in place of marriage and paid but fleeting attention to child–adult relations. Indeed he spent a considerable amount of time on theorizing different forms of love in intimate relationships in late modernity.

Giddens provided a social theory about these shifting patterns that had been subject to frequent political debate and controversy under the economic or neoliberal regimes of Thatcher and Major. His social theory is built upon the various critical feminist and radical social theories that had emerged and proliferated during the previous three decades of social movements and social liberalism (David, 2003). These had sought to transform traditional family lives through political projects of social change and emancipation. At the same time sociologists and feminists had sought to theorize these changes in more conventional academic ways. The patterns and trends in social transformation of family lives became the main subject of considerable sociological analysis and debate. Feminists and sociologists highlighted various facets of the changes in family lives and work, such as theorizing the shifts toward more personal public agendas (Jamieson, 1998; Fraser, 1997). Their theoretical and methodological concepts also took account of more personal agendas about family life changes including what has become known as "the autobiographical turn" in the social sciences (Chamberlayne et al., 2000).

"Personal Responsibilities" as the New Policy Discourse for Changing Family Lives

It was argued that changing family forms were intimately connected with the wider economic and political contexts, and also to transformations in social networks and forms of social capital as well as economic and cultural capital (Bourdieu, 1992; Edwards, 2002). Both political commentators and social researchers as academics identified and theorized the changes, but from a variety of contested perspectives. For instance, a guru of neoliberalism and the New Democrats in the United States, Amitai Etzioni, who at the time was also president of the American Sociological Association propounded a new approach that he called "communitarianism" (Etzioni, 1992). This particular approach required the development and renewal of local, regional, and neighborhood communities, through voluntary effort rather than political investment. Whilst the language of communitarianism remained gender-neutral, a specific sexual and social division of labor was both envisaged and specified. This entailed the key notion of "the parenting deficit" for dependent and young children reared in communities in which parents, or rather mothers were involved in paid employment whilst their children were little and young. These ideas appealed to New Democrats in the United States and to New Labour in opposition in Britain. The ideas were publicized by the IPPR and given credence in official New Labour policy documentation.

One key aspect of the neoliberal political project had been to try to reverse the trends in changing family lives (Murray, 1994, 1996). In particular, the growth of lone-parent families with dependent children created either through divorce or cohabitation, and a "refusal" to marry was the main target of concern to neoliberal politicians in Britain (Kiernan et al., 1998). There was a similar debate in the United States about transformations in family lives and their impacts upon fiscal and social welfare policies. Moreover, the British conservative political debate borrowed from the United States the notion of "a moral panic" about the demise of the traditional nuclear family, especially amongst socially and economically disadvantaged people. The key problem was deemed to be the emergent new forms of family, consisting of single-mother households, or what were called female-headed households in the United States (Murray, 1994, 1996).

In the political debates in the United States such single-mother families were usually associated with race and ethnicity as indicators of poverty whilst in Britain the debate hinged upon economic disadvantage. In both the United States and in Britain, whatever the discursive rhetoric and characterization, these mothers in such families were demonized for their behavior (Arnot et al., 1999; Standing, 1999). The political debates tended to focus on issues in relation to social welfare and the fiscal problems of providing public support for such families. Thus the emphasis was on finding more appropriate social and public policies to deal with the growing crisis of social welfare, especially for young and single mothers and their dependent children. However, little attention was paid to the implications of changing forms of family and parenthood for education, schools, and families. (This is a point to which I will return.)

For instance in the United States both Republican and Democratic regimes, characterized as conservative economic or neoliberal regimes, focused upon revised political solutions to the problems of diverse family forms. In particular, they targeted female-headed households as subjects for reform since they lacked traditional patriarchal family forms. One key instance of this was the welfare reforms of the 1990s, enacted eventually by a very conservative Congress, whilst the president, Bill Clinton, was a New Democrat. These welfare reforms abolished traditional forms of social welfare (known as Aid to Families with Dependent Children—AFDC) and introduced Temporary Aid for Needy Families (TANF). Fraser and Gordon (1997) provided an excellent feminist critique of the concept of dependency entailed in social welfare, although they focused on its problematic nature for the single mothers, rather than their own economic and socially dependent children. The transformation from AFDC to TANF created even more serious problems of economic and social dependency for such women and their children, given the shifts toward temporary and limited forms of economic support. Moreover, the new language of "needy families" also created more discretion in how it might be defined (Schram, 2000). A key concept introduced by this legislation was that of "*personal responsibility*," defined in terms of single women's economic and social responsibility for their dependent children, rather than their traditional reliance on forms of social welfare.

In Britain similar measures to revise and reform social welfare were introduced first under the Conservative governments to try to stem the growth of lone-parent families becoming dependent upon social welfare (Bradshaw and Millar, 1991). Second, New Labour adopted and adapted these measures, whilst recognizing the irreversible character of these kinds of social and familial changes and re-designated the issues about social exclusion and inclusion. Indeed, New Labour sought to reinforce the labor force participation of women in a variety of family circumstances, including with dependent children, rather than continuing the social democratic approach of providing forms of economic and social protection for such mothers. Indeed, a key component of the social, political, and cultural turn has been the recognition of the necessity and centrality of paid work over other social and personal responsibilities. Notions of the active and responsible citizen (as an adult responsible for personal and social welfare) slowly also began to enter the arena and lexicon of New Labour, as an antidote to social exclusion.

Thus the debates raged about the specificity of these changing social, sexual, and cultural forms and their impacts and effects on both policies and politics and upon changing social and sexual identities. Indeed an aspect of the political discourses and attempts to reform social welfare was to try to reverse transformations in family lives and recreate traditional family forms. However, where such families had only females as heads the socially liberal principle of providing economic and social support for both women and children was modified to ensure "personal responsibilities" rather than state aid, returning to Victorian family values (Arnot et al., 1999). A more subtle approach to these matters as part of the cultural turn was the attempt to develop not only fiscal measures but also new educational strategies around these ideas. These transformed traditional perspectives on education as well as on social welfare had implications for gender and generational relations.

Traditional Patterns of the Family-Education Couple

Notions of the partnership between families and education, particularly as mediated by the state, had entered the lexicon of public service and educational provision in the postwar era of social democracy and liberalism. The particular notion of "the family-education couple" was one coined by the French Marxist philosopher Louis Althusser, who argued that not only was it important as part of the social and political structure but that its significance lay in its ideological effects. He thus conceptualized "the family-education couple" as an ideological state apparatus (1971). I used this conceptual social analysis, together with a feminist perspective in an attempt to elucidate the relations between the state, the family, and education in the late 1970s (David, 1980). The notion of the couple, however, was implicitly if not explicitly predicated upon traditional family patterns of marriage, rather than a more loose form of partnership or relationship outside of wedlock and one in which there was more equality between partners than based upon a traditional sexual division of labor. It was possible to conceive of traditional forms of state educational provision as forms of patriarchal control, namely with the state through either federal or state government as "the male provider" of education. The roles of families and communities in such educational provision could be characterized relatively limited to a female dependency position.

Whilst these patterns took a variety of complex forms under social democracy in Britain and social liberalism in the United States, families and parents tended to be provided with a limited and not very strong role, one that I once characterized as of "reciprocal inequality" (David, 1977). In the United States however a semblance of community and parental participation was in greater evidence in some small communities on the east coast and in New England committed to community control (Fein, 1971; Levin 1970). In a very small number of communities in New England there remained some instances of Grecian forms of direct democracy, such as the townships of Wellesley and Weston, than in countries more ideologically committed to social democracy and socialism (David, 1975). However, these were largely exceptions to the more traditional bureaucratic principle of city, state, and federal government in education.

Although most commentaries and academic studies of such forms of educational control in the 1960s–1980s tended to ignore the gender and sexual dimensions, there were one or two emergent studies by feminist academics and sociologists who pointed to these gendered and sexist patterns (Stambach and David, in press). Studies by women began to form the early material for the emergence of what became known first as the women's liberation movement and later as "second wave" feminism to distinguish it from the "first wave" of suffragettes. For example, Betty Friedan (1963) in her much acclaimed feminist critique of suburban communities in the United States, *The Feminine Mystique*, pointed to the sex segregated roles of mothers and fathers in relation to child rearing and schools. In describing "the problem that has no name" she highlighted in particular the frustration that many suburban housewives felt about being confined to home, family, and domesticity and their role in being supportive of their children and schools, whilst the husband-fathers developed and progressed in their careers. This study was a clarion call to white middle-class housewives to develop themselves and obtain more education to move out of these frustrating situations and return to careers and work. Her call was highly successful in inaugurating a strand of the women's movement in the United States along with the campaigns of other liberal feminists (Stambach and David, in press). However, she herself reneged upon these perspectives in her later work, arguing for a more traditional approach to family lives (Friedan, 1982).

Nevertheless, she identified a key issue that has become the subject of much more nuanced social analysis and educational research, the roles of parents in relation to schools and education and in relation to changing family and social patterns and community development. Numerous studies subsequently in both the United States and Britain have highlighted these issues amongst diverse social class and ethnic minority communities in the context of shifting expectations and changing social and familial patterns (David, 1993). Thus there has been a mushrooming of social and educational research on the issues of family education or home and school relationships, that have employed the analogies with traditional marriages and forms of social expectations (David, 1993; David et al., 1993). Although the notion of a sexual division of labor in marriage might have been implicit in these studies much of the research has not highlighted the ways in which sex and gender have been threaded through these various patterns of relationships.

However, at the same time, over the last 30 years, it has become more normative for parents to be enjoined to become involved in education either at home or at school. And more importantly, it has become a much greater normative expectation, implicit if not explicit that much of that involvement would entail mothers rather than fathers. A traditional nuclear family relationship (and possibly also suburban) was envisaged and expected. However, a major social and economic change over this period has been mothers' patterns of paid employment and their involvement in varying forms of work for women as mothers. Thus the patterns of parental involvement in various educational processes have become more complex yet largely predicated on a middle-class model, and one essentially of white traditional nuclear families. Thus families not conforming to these patterns have tended to have had opprobrium heaped upon them. Indeed, some of the classic typologies of parents and education have built into them these normative notions (Epstein, 1992).

Educational Partnerships: Sex, Gender, and Generation

The changing forms of governance of education began to take place during the high point of conservative regimes, with the shift from state and bureaucratic patterns of control toward the introduction of market forces and privatization. Under traditional social democratic or liberal regimes the state had played the major role in the design and implementation of educational policies, albeit with differing roles for levels of government. Neoliberal regimes initially provided critiques of the stultifying nature of these forms of control, which did not allow for choice or opportunities, and thus provided mediocre forms of education. Most importantly it was argued that parents and families were not provided with a part in educational provision, but educational professionals played the key part in educational decision-making.

The introduction of business forms and methods was aimed at increasing both choice and standards, and above all, excellence in education (Whitty et al., 1998; Witte, 2000). New measures of performance were also to be introduced to ensure these new standards. These new approaches in Britain have often been entitled "new managerialism" (Clarke and Newman, 1997). Under New Labour in Britain ideological commitments to forms of educational communities began to shift slowly. It was a similar process to the ways the shifts under economic liberalism were taking place, namely more social transformations. But key was the shift to the public discourse of partnerships and about balancing "stakeholders" in education. This was most usually predicated on the notion of business partnerships, with little recognition of the more nuanced form of partnerships, based around sexual relations. However, Fairclough (2000) had noted the discursive shifts in New Labour's lexicon.

Thus the political discourses of both New Labour and the New Democrats in the United States acknowledged and recognized the wider global and social transformations and indeed used the new language of globalization (Fairclough, 2000). The political discourses also recognized how they were also constituted by the discourses. Moreover, their discourses reflected and were reflective of these shifting notions toward new notions of sexual partnerships through and within education. Two interlocking and interrelated discourses both framed within this new language or discourse of sexual partnerships slowly emerged. One was about the changing general framework of relations between new generations of families and schools and the other was about the pedagogies and practices of new sexual relations within schools. Both were highly contested areas, and I move on now to illustrate these shifting notions within educational discourses by looking first at the frameworks and second at pedagogies and practices.

Educational Partnerships and Transforming Generations of Female-Headed Families

Educational partnerships as a notion were very diverse, but a strong notion was linked with developing a new framework for schools and relations with families as part of a wider, global and changing community and issues of social exclusion. Thus there was clear acknowledgment of the complexity of social and family changes and the need to renew and renegotiate relationships as school was deemed to require changes in relation to wider social and technological changes. Family, gender, social, and educational changes were intimately interwoven in complex ways. Education and the changing knowledge economy all became parts of these reconstituted notions as did the additional notion of stakeholders, which had been initially developed at the height of neoliberalism. In particular, such partnerships under neoliberal regimes became linked with discourses about social capital and communitarianism, as we have noted. Social networks as one aspect of social capital were seen as "resources the community can offer schools" (Smit et al., 1999) as well as economic markets and business orientation as other more conventional aspects of partnerships.

Thus the rather traditional notions of home–school relations and parental involvement in education, either at home or at school, also were reconsidered and revised to take account of wider social and educational changes. Whereas in the past families and parents were enjoined to participate in their children's education "for the sake of the children," on a middle-class model, such notions of participation and involvement were reconstituted to the wider notion of partnerships. In part this was very much part of a bigger and more global process of internationalization. Communities at both the local and regional or state level were deemed to be changing along with educational changes, affecting generations of families and particularly young people reared in a diversity of family, ethnic, and social contexts. These myriad changes affected notions of education in relation to family and community and at the same time people's perspectives and understandings of family and community and their own gender and sexual identities. These were slowly more articulated within social and educational research, in particular as the basis for policy and educational reforms.

At the same time notions of family choice of education also became more closely and clearly specified, such as through new more market-oriented but community approaches and options. In the United States the refinement of notions of both vouchers and charters for diverse and closely associated groups of parents to set up and run their own schools in partial partnership with the local school board, mushroomed in the late 1990s (Yancey, 2000). In particular, these new forms of educational frameworks for the delivery and implementation of new forms of schooling made space for far greater parental participation through communities as well as families. Although not always acknowledged as such the spaces created in these newly defined school communities and organizations were usually for women's greater involvement in the processes as professional educators as well as parents (Yancey, 2000; reviewed by Stambach et al., 2002). Shifts from community control to parental control were quite dramatic amongst these forms of alternative schooling and seen as part of a newly emergent postmodern society.

To some extent the discourse of partnerships has been replacing the discourse of stakeholders within education, with an implied and implicit if not explicit notion of far greater relative equality between partners. The discourse of stakeholders (Hutton, 1995) was coined in the heyday of economic liberalism by conservative thinkers and did not bear with it any notion of relative equity between stakeholders, but broadened the canvass to include business and community participants with economic investments in education. The discourse of partnerships has adapted and modified the notion of stakeholders and implied greater equity between participants and more gender equity. In the following, I offer instances of the complexities of the changes, especially in relation to gender and generation.

A critical and key instance of the ways these concepts were globalized can be found in the reports of European projects linked to U.S. projects for education, particularly in Eastern Europe. For example the European Research Network Association of Parents and Education (ERNAPE) brought together scholars and researchers around international projects about the involvement of parents in education. The flyleaves of the two most recently edited conference proceedings illustrated the ways in which this discursive rhetoric became central. For the edited volume for the first conference (Smit et al., 1999), about countries all over the world and entitled *Building Bridges Between Home and School* the cover jacket stated, "the forms of involvement run from orientation to *partnerships* in specific subjects to systems, *models and strategies for partnerships*" (my emphasis). The more recent conference proceedings volume had a very similar refrain (Smit et al., 2001). "The different forms of involvement run from orientation to *partnerships* in specific subjects to systems, models and strategies for *school-family-community partnerships*... cooperation between the various *partners* in the collaboration between parents, schools and communities" (my emphasis).

In Britain, under New Labour these notions became crucial to the revisions to the educational framework. Thus New Labour came to power with a commitment to what Prime Minister Tony Blair called "Education, Education, Education" and redefined the whole approach as part of the modernization project and the Third Way (Giddens, 1998) as we have already noted. Indeed its very first piece of social legislation was an educational measure to revise school relations. Entitled the 1998 School Standards and Framework Act it set out not only new perspectives on school improvement and achievement but, perhaps more importantly for this essay, revised relations between schools, communities, and families, through redefining home and school relation. Through home–school agreements (formerly contracts) came a new definition of partnership of relative or reciprocal inequality. These agreements were set to become mandatory and required each school to set out its responsibilities to families and pupils and the commensurate and reciprocal parental responsibilities for/to schools on behalf of their children as pupils. The model agreement set out in the advice to schools and governors made explicit that parental responsibilities were greater than those of schools but the claim was of a partnership between the various stakeholders and families and schools in particular. These were reinforced and reemphasized in subsequent legislation and especially the Learning and Skills Act 2000 (Monk, 2001).

These notions of new forms of partnerships are replete with double entendres around sexual relations although in terms of parent–child relations partnerships have become much more punitive. As we have already noted in the United States there are instances of the ways in which notions of partnerships built upon changing social and community participation and family changes to open up spaces for more women's participation as professionals, parents, and community organizers and activists. These all build upon women's diverse and complex changing employment and community activities, as both parents and also as women have increasingly entered the labor market, especially the education and social service labor markets.

Following Ehrenreich (1990) who showed how the middle classes in the United States feared for their children's lack of educational progress and social mobility, Vincent and Ball (2002) have demonstrated the ways in which families and mothers are now expected to make choices and be involved in their children's education. There is similar evidence for Britain to suggest that women as mothers and as workers within education and childcare take their new expectations as partners and participants very seriously indeed. Vincent and Ball (2002) have shown how dramatic these shifts have been for the parents, and for mothers especially of very young children at preschool and in early childhood. Thus parents, or mothers, become active partners and participants in the educational process, from shortly after the birth of their children and within both informal and formal early childhood education, enticed and exploited by new forms of advertising and business activities.

Similarly both Reay (2002) and Crozier (2002) have also illustrated how mothers in a diversity of family and social circumstances have redefined their partnerships with education away from traditional notions of the collective or common good toward a more individualized notion for their children. A stark instance of these developments has been the accumulating evidence of the extent of private tuition and tutorships to bolster and extend schooling for young people. Thus mothers have increasingly engaged with education in more individualized and diverse forms in seeking to increase their children's educational life chances, opportunities, and circumstances. In the process, these have come to be redefined as forms of active and responsible citizenship and new forms of partnerships.

More complexly, and yet clearly also associated with wider family, social, and educational transformations, mothers' responsibilities as parents have been adapted and extended, not only as active partners to ensure the effectiveness of schooling and the wider educational processes, but also where their children are not participating. There were two recent *cause celebres* in the media in Britain that demonstrated the shifting discourses around parental responsibilities for young people at risk of social exclusion through school nonattendance, hitherto deemed to be school truants. These illustrated what has been called new forms of risk management (Bullen et al., 2000). Historically legal definitions of parental responsibility for truancy had entailed fathers rather than mothers being held legally liable (David, 1980). Although fathers had been prosecuted the usual punishment had been a fine rather than imprisonment. Historically concern had centered upon boys' truancy and nonattendance rather than girls, and linked to male forms of juvenile delinquency. In a complex revisioning of these patterns the changes centered upon mothers and daughters rather than fathers and sons.

Illustrating the shifting patterns of parental relationships and responsibility that had also been redefined through various child care and educational laws (David, 1993; Monk, 2001), two legal cases focused upon mothers' responsibilities and changing patterns of legal penalties. In the first case, the single, white working-class mother of two teenage girls, aged 15 and 13 years old, was sent to prison as a result of a school nonattendance order. This order had been served on her after numerous attempts to try to get her daughters to attend school. She had not taken proper parental responsibilities for her daughters' education and had allowed them to stay at home rather than attend compulsory schooling. Arguments also focused upon her own mother's death, which had triggered trauma and inability to cope as the maternal grandmother had done all the work of caring for the daughters, and forms also of drug abuse. However complex the details of the actual legal case, it was important to note the recognition and reinforcement of female and maternal responsibilities. Not only, however, did public concern focus upon maternal personal responsibilities but also punitive measures were invoked, uniquely, to ensure that young women continue to attend school to ensure habits of work and employment opportunities.

The second media "case" exhibited very similar attributes and helped to consolidate a new form of "risk management" (Bullen et al., 2000) for women as both mothers and daughters in situations of social exclusion. In this case a single non-married mother of a daughter aged 12 years old was also threatened with prison because of her daughter's nonattendance at school. The daughter was defined as being school phobic and afraid to attend school because of threats of bullying. The case was to be brought by a male headteacher concerned about the mother's lack of exercise of her parental responsibilities and lack of sympathy for issues of school phobia as defining school and social exclusion. Although in this particular instance the father was mentioned, he was not married to the mother and although deemed to be her partner was not held legally responsible. In such circumstances, in order for non-married male partners to be given and assume parental responsibilities for their children's schooling they had to apply for a parenting order under the Children's Act of 1990.

What was the significance of gender in these two cases? The first one was the case of a mother who had "cracked" up on her own mother's death but was clearly living in what had been defined as a socially excluded family situation. Similarly the second one also illustrated how the notions of social exclusion, parental responsibility, and personal responsibility were being designated within new social and cultural contexts. Thus gender had indeed become significant in the changing forms of family and community, with single mothers held personally responsible economically and socially for the education of their daughters. Moreover, notions of partnerships have become more complex in such situations of educational and social transformations, whereby male partners have become either absent or shadowy figures in educational responsibilities. Women as mothers have been enjoined to take on these traditional male responsibilities in situations where men have been released from such responsibilities. Moreover, such women are now expected to shoulder both economic and regular responsibilities for their children's education in nontraditional ways.

The three daughters who were the subjects of these two legal cases also highlight the revised significance of gender. They were of the age where they could also have been seen to be problematic as they were themselves school-age mothers or part of the group problematized as crucially at risk of teenage pregnancy. I turn now to how these issues interlock with familial transformations and are underpinned by complex and complicated changing notions of partnerships in education.

Sex and Sexuality Education: New Partnerships or Old Marriages?

Under neoliberal regimes there has also been a major debate about not only how to deal with family changes in terms of personal and parental responsibilities but also how to teach new generations of young people about these changing forms of sexual relationships and partnerships (Epstein and Johnson, 1998). Recognizing such changing family relationships and their implications for new families, new relationships and new responsibilities emerged slowly in relation to welfare reforms in the United States and the British reforms of social welfare. However, the political debate about changing forms of sexual and social relationships and their specifically educational implications originated in the United States under a complex set of political circumstances in the mid-1990s (Landry, 2002). In Britain it emerged strongly under New Labour, borrowing from the United States where the reproblematizing of the issues was linked with both the modernizing of the social and the remoralizing of social welfare (Carrabine, 2002).

However, in both the United States and Britain the issues were not specified as general but particular to issues of the risks of social exclusion. As noted earlier, the discursive solutions centered upon new forms of "risk management." In the United States the moral and religious right emerged as crucial to defining the terms of a new educational agenda for new sexual identities and relationships. They identified changing patterns of marriage and family relationships, whereby over 50 percent of families with dependent children were not in traditional marriages and linked these with the problems of teenage pregnancy. In the United States, by the mid-1990s, rates of teenage pregnancy were identified as being the highest in the Western world.

A simple and unproblematic and yet new "educational" solution, related to the religious right and the law was invoked, namely abstinence-before-marriage education (David, 2003). This became tied to federal funding initiatives and states were only given federal funds for educational programs if they committed themselves to developing such sex education programs. This scheme began in 1996 and was further developed by the Bush administration in refusing to endorse programs of sex education for international aid if they did not conform to such rigid stereotypes. Thus the United States accepted the necessity of education and public discourses about private and intimate matters, and yet engaged only with very traditional moralistic solutions. As Landry (2002, p. 12) put it:

> Most in the abstinence-until-marriage movement vaguely portray the period of abstinence as ending sometime shortly after the teen years. But in the US, the age of first marriage has been steadily rising since the 1970s. In 2000, the age at which half of men and women first married (or the median age at marriage) was 25.1 and 26.8 respectively. Moreover, a significant proportion of men and women remain unmarried into their thirties. For instance, by age 30.9, 25 per cent of women in the US have still not married.
>
> While it may be safer to marry a partner who has never had sex, the effort to find one may further delay age at marriage. The abstinence-until-marriage movement simply fails to take into account the scope of sexual activity in our society. Currently the median age of first intercourse in the US is 16.9 for men and 17.4 for women. By 21 years of age, 90 per cent of men and women have had sexual intercourse. In fact, only 22 per cent of women aged 18–59 and 10 per cent of men waited until they were married to first have intercourse.

In Britain, New Labour's response was altogether more complexly related to social and family transformations, whilst at the same time being underpinned by concerns about rising rates of teenage pregnancy as central. Thus there was a political recognition of changing social lives and sexual relationships or partnerships but the question of how to respond in educational terms was more complicated. New Labour set up an administrative department—the Social Exclusion Unit—at the heart of government on coming to office in 1997. This unit was charged first with the question of investigating teenage pregnancy and its first report was on this topic (Social Exclusion Unit, 1999). This report identified the problems of teenage pregnancy in Britain as being the highest in Western Europe (and second only to the United States). However, it did not follow the moralizing and stigmatizing rhetoric of previous British or the U.S. administrations about out-of-wedlock teenage pregnancies.

New Labour identified the problems as a result of and at the same time a cause of social exclusion (Carrabine, 2002) and as linked to wider social and family transformations, and shifted the responsibilities from social neglect to government for a response. The response was also linked to the new knowledge economy. Thus it identified these changing sexual and family relationships as the result of lack of knowledge and information, and the mixed messages associated with a more information and media oriented society. It also identified forms of risk management for young people through knowledge acquisition and learning about social and sexual risks, as well as learning about sexual relationships. It also specified an educational program for young people to learn about these kinds of issues including acquiring information about social welfare, social supports, and active citizenship responsibilities.

Central to this program was the idea of sex and relationship education (SRE) linked with citizenship education, also a new program, and both requiring a new pedagogy. A year after the report (Social Exclusion Unit, 1999) was published, national guidance on sex and relationship education was also published for the use of schools and governing bodies. This covered three elements to constitute a good program of SRE, namely attitudes and values, personal and social skills, and knowledge and understanding. Although it was recognized that the multitude of social and family changes that had taken place over the previous two to three decades required new forms of education, an underpinning assumption remained of the centrality of marriage or serious and committed sexual relationships or partnerships. However, the revised program also developed as part of assumptions about sexualized culture and the need to educate about the various risks, associated with the changes, health, social, and sexual.

Thus, both Britain and the United States recognized the centrality of changing family and personal relationships to wider changing community and educational partnerships, including pedagogies for sexuality and sex education. Yet little attention had been given to the underpinning discursive practices, nor to the gender and generational dimensions of these social transformations. For Britain, however, these transformations were closely and clearly linked with what had been termed the personal and cultural turn, and as part of New Labour's modernization project. Thus notions of educational partnerships, reliant upon sexual and gender changes represented part of the Third Way, whereby intimate, sexual relationships or partnerships constituted and were constituted by education.

Conclusions

Under neoliberal regimes there have been shifting discourses in policy and practice around partnerships that have mapped on to shifts in social and sexual transformations and constituted part of the modernization of these regimes. New generations of families have emerged over the last two decades in which traditional patterns of relationships, partnerships, and marriages have been transformed and are transformative of policy changes (Duncan and Edwards, 1997, 1999). Moreover, personal lives and responsibilities have become more under the public gaze, as have these very social and family transformations. Thus not only welfare reforms but also educational reforms have been implicated in these social and policy transformations. Modernizing both the social and the personal has been a key constitutive element.

Instances were offered of how the discursive shifts that developed notions of educational partnerships could arguably be linked around changing families. There were especially gender dimensions to the ways in which the re-specification of not only the social but also the educational implicated women as mothers and professionals. New programs of community and civic involvement and engagement around educational partnerships implicitly entailed recognition of mothers in relation to their children's schooling. Both the educational reforms of the frameworks for community engagement and participation are built upon traditional patterns of a sexual division of labor, and an implicit recognition of women's educational and emotional labor, in pursuit of their children's education. Thus their identities have been reconstituted as both mothers and daughters.

Perhaps more importantly from the perspectives of the emergent reforms around social exclusion and the reforms of social welfare have been how the revised programs have implicated particular groups of mothers. Thus instances were also offered of how the legal frameworks have become more punitive toward women as mothers. Moreover, it appears that as part of the more global social transformations the expectations that women as mothers take more personal responsibilities both financially and educationally. These also have far-reaching implications for future generations as their children—their daughters especially have been pulled into more responsibilities through their mothers' emotional labor as well as educational work. In also re-moralizing the social and welfare women's personal responsibilities have come under a new gaze. Here particular groups of socially excluded women have had their identities reconstituted through these policies and practices.

There are far more diverse patterns of partnerships and relationships, particularly outside of traditional marriages and in particular ethnic, cultural, and social class communities. But some may be summarized as bifurcated strands of new generations of families. There are on the one hand, both parents and children/young people who may become parents while young: at school-age or at least as teenagers. A moral panic emerged about these new generations of young people, their identities and perspectives, of how through a changing culture they were being sexualized earlier than in previous generations. These groups were identified in both the United States and Britain as constituting those at risk of social exclusion. Thus educational programs were devised to counteract these changing generations and provide a greater knowledge base for them to consider forms of risk management.

On the other hand, a major trend has been toward the majority of families having children later in nontraditional partnerships and outside of marriage. In many of these families marriage if the partners plan on it, often occurs after the children are in school. These patterns of nontraditional partnerships have reached such proportions that they have occurred not just in working- or under-class but also in middle-class families. Sexuality and relationship education was devised not only to deal with risk management for groups potentially at risk of social exclusion but also as an expression of a more sexualized and personalized culture. Thus shifts in partnerships and relationships outside traditional marriage have constituted and been constituted by wider social and cultural transformations.

These broad shifts in educational partnerships thus give expression to the cultural and personal turn. They also constitute the ways in which the modernization project of neoliberal regimes has transformed and been transformed by global family, gender, and sexual shifts and changes. In education and welfare reforms, personal responsibilities and transformations especially in women's personal responsibilities for their children and themselves as parents, mothers especially have been critical and crucial. They have engendered key transformations in sexual and social justice, such that child care and education no longer are recognized as key but rather personal responsibilities are critical to re-moralizing and modernizing the social.

References

Althusser, L. (1971). "Ideology and Ideological State Apparatuses." In *Lenin and Philosophy and Other Essays*. London: New Left Books.
Arnot, M., David, M.E., and Weiner, G. (1999). *Closing the Gender Gap: Post-War Education and Social Change*. Cambridge: Polity Press.
Blair, A. (1998). *The Third Way: New Politics for the New Century*. London: Fabian Society pamphlet.
Bourdiew, P. (1992). *Language and Symbolic Power*. Cambridge: Polity Press.
Bradshaw, J. and Millar, J. (1991). *Lone Parents in the UK*. Dept of Social Security Research Report, No. 61, London: HMSO.
Bullen, E., Kenway, J., and Hey, V. (2000). "New Labour, Social Exclusion and Educational Risk Management: The Case of 'Gymslip Mums.'" *British Educational Research Journal*, 26, 441–451.
Carrabine, J. (2002). "New Data Old Stories: Remoralizing Teenage Pregnancy Under New Labour." Paper presented at the Gender, Law, and Sexuality Conference, Keele University.
Chamberlayne, P., Bornat J., and Wengraf, T. (Eds.) (2000). *The Turn to Biographical Methods in the Social Sciences*. London: Routledge.
Clarke, J. and Newman, J. (Eds.) (1997). *The Managerial State*. London: Sage.
Commission on Social Justice (1994). *Social Justice*. London: Institute for Public Policy Research.
Crozier, G. (June 2002). Beyond the Call of Duty: The Impact of Racism on Black Parents' Involvement in Their Children's Education. Paper presented at the market Forces Seminar Group, Kings College, London.
David, M.E. (1975). *School Rule in the USA*. Cambridge, Mass.: Ballinger Publishing Co.
David, M.E. (1977). *Reform, Reaction and Resources: The 3 Rs of Educational Planning*. Slough, Berks: National Foundation for Educational Research Publishing Company.
David, M.E. (1980). *The State, the Family, and Education*. London: Routledge and Kegan Paul.
David, M.E. (1993). *Parents, Gender, and Education Reform*. Cambridge: Polity Press.
David M.E. (1998). "Education, Education, Education." In H. Jones, and S. McGregor (Eds.), *Social Issues and Party Politics* (pp. 74–91). London: Routledge Labour
David, M.E. (1999). "Home, Work, Families and Children: New Labor, New Directions and New Dilemmas." In *International Studies in the Sociology of Education*, Vol. 9, part 3, pp. 209–229.
David, M.E. (2002). *Personal and Political: Feminisms, Sociology and Family Lives*. Cambridge: Polity Press.
David, M.E. (2003). "Teenage Parenthood is Bad for Parents and Children: A Feminist Critique of Family, Education, and Social Welfare Policies and Practices." In M.N. Bloch and T.S. Popkewitz (Eds.), *Governing the Child, Families and Education: Restructuring the Welfare State* (pp. 149–171). New York: Palgrave.

David M.E. (2003a). *Personal and Political: Feminisms, Sociology and Family Lives*. Cambridge, England: Polity Press.
David M.E. (2003b). "'Teenage Parenthood is Bad for Parents and Children': A Feminist Critique of Family, Education and Social Welfare Policies and Practices." In M. Bloch and T.S. Popkewitz (Eds.), *Restructuring the Governing Patterns of the Child, Education and the Welfare State* (New York and London: Palgrave).
David, Me., Edwards, R., Hughes, M., and Ribbens, J. (1993). *Mothers and Education: Inside Out? Exploring Family-Education Policy and Experience*. London: Macmillan.
Duncan, S. and Edwards, R. (Eds.) (1997). *Single Mothers in International Context: Mothers or Workers?* London: UCL Press.
Duncan, S. and Edwards, R. (1999). *Lone Mothers, Paid Work, and Gendered Moral Rationalities*. London: Macmillan Press.
Edwards, R. (January 2002). *Presentation to the Inaugural Seminar of ESRC Families and Social Capital Research Group*. London: South Bank University.
Ehrenreich, B. (1990). *Fear of Falling: The New Middle Classes in the USA*. New York: Basic Books.
Epstein, D. and Johnson, R. (1998). *Schooling Sexualities*. Buckingham: Open University Press.
Epstein, J.L. (1992). "School-Family Partnerships." In M. Alkin (Ed.), *Encyclopedia of Educational Research*, 1139–1151. New York: Macmillan.
Etzioni, A. (1992). *The Parenting Deficit*. London: Institute for Public Policy Research.
Fairclough, N. (2000). *New Labour, New Language?* London: Routledge.
Fein, L. (1971). *The Ecology of Public Schools*. New York: Pegasus.
Fraser, N. (1997). *Justice Interruptus: Critical Reflections on the Postsocialist Condition*. London: Routledge.
Fraser, N. (2000). "Radical Academia, Critical Theory, and Transformative Politics: An Interview with Nancy Fraser," *Imprints*, 4, 197–212.
Fraser, N. and Gordon, L. (1997). "A Geneology of Dependency: Tracing a Keyword of the U.S. Welfare State." in Fraser, N. (Ed.), *Justice Interruptus: Critical Reflections on the Postsocialist Condition*. London: Routledge.
Friedan, B. (1963). *The Feminine Mystique*. Harmondsworth, UK: Penguin.
Friedan, B. (1982). *The Second Stage*. London: Jonathan Cape.
Giddens, A. (1992). *The Transformation of Intimacy: Sexuality, Love, and Eroticism in Modern Societies*. Cambridge, England: Polity Press.
Giddens, A. (1994). *Beyond Left and Right*. Cambridge, England: Polity Press.
Giddens, A. (1998). *The Third Way: The Renewal of Social Democracy*. Cambridge: Polity Press.
Giddens, A. (1999). *Runaway World*. Cambridge: Polity Press.
Hutton, W. (1995). *The State We're In*. London: Verso.
Jamieson, L. (1998). *Intimacy: Personal Relationships in Modern Society*. Cambridge: Polity Press.
Kiernan, K., Land, H., and Lewis, J. (1998). *Lone Motherhood in Twentieth Century Britain*. Oxford: Clarendon Press.

Landry, D.J. (2002). "Sex Education. A View from the United States." In *Teenage Sex: What Should Schools Teach Children?* London: Institute of Ideas, in collaboration with Hodder and Stoughton.
Leach, Sir. E. (1968). *A Runaway World: The Reith Lectures.* London: British Broadcasting Corporation.
Levin, H. (Ed.) (1970). *Community Control of Schools.* New York: Simon and Schuster.
Levitas, R. (1998). *The Inclusive Society? Social Exclusion and New Labour.* London: Macmillan.
Monk, Daniel (2001). "New Guidance/Old Problems: Recent Developments in Sex Education." *Journal of Social Welfare and Family Law,* 23, 3, 271–291.
Murray, C. (1994). *Underclass: The Crisis Deepens.* Choice in Welfare No. 20. London: Institute of Economic Affairs Health and Welfare Unit and The Sunday Times.
Murray, C. (1996). *Charles Murray and the Underclass: The Developing Debate: "The Emerging British Underclass" and Underclass: the Crisis Deepens.* Choice in Welfare No. 33. London: Institute of Economic Affairs and The Sunday Times.
Reay, D. (June 2002). *Exclusivity, Exclusion and Social Class in Urban Educational Markets.* Paper presented to the Market Forces Seminar Group. London: Kings College London.
Schram, S.F. (2000). *After Welfare. The Culture of Postindustrial Social Policy.* New York: New York University Press.
Silver, H. (1994). *Good Schools, Effective Schools and Judgements, and Their Histories.* London: Cassell.
Smit, F., Moerel, H., van der Wolf, K., and Sleegers, P. (Eds.) (1999). *Building Bridges Between Home and School.* Institute for Applied Social Sciences, University Nijmegen, Netherlands.
Smit, F., van der Wolf, K., and Sleegers, P. (Eds.) (2001). *A Bridge to the Future: Collaboration Between Parents, Schools and Communities.* Institute for Applied Social Sciences, University Nijmegen, Netherlands.
Social Exclusion Unit (1999). *Teenage Pregnancy.* June Cm 4342. http://www.cabinet-office.gov.uk/seu/1999.
Stambach A. and David, M.E. (in press). "Feminist Theory and Educational Choice: How Gender has been Involved in Home-School Choice Debates." Signs: *Journal of Women and culture.*
Standing, K. (1999). "Lone Mothers and Parental Involvement: A Contradiction in Policy?" *Journal of Social Policy,* 28, part 3, 479–517.
Vincent C. and Ball, S.J. (June 2002). *Social Reproduction is a Risky Business.* Paper presented at the Market Forces Seminar Group. London: Kings College, London.
Whitty, G., Power, S., and Halpin, D. (1998). *Devolution and Choice in Education: The School, the State, and the Market.* Philadelphia: Open University Press.
Whitty, G. (2002). *Making Sense of Educational Policy.* London: Paul Chapman, Publishing.

Witte, J.F. (2000). *The Market Approach to Education: An Analysis of America's First Voucher Program*. Princeton, N.J.: Princeton University Press.

Yancey, P. (2000). *Parents Founding Charter Schools: Dilemmas of Empowerment and Decentralization*. New York: Peter Lang.

Yancey, P. (2000). *Parents Founding Charter Schools: Dilemmas of Empowerment and Decentralization*. New York: Peter Lang; and see also Review Symposium in *British Journal of Sociology of Education*, 2002 Vol. 23, No. 1 pp. 123–133, with reviews by Stambach, Bush, Fitz, and David).

CHAPTER 9

PARTNERING TO SERVE AND SAVE THE
CHILD WITH *POTENTIAL*: REEXAMINING
SALVATION NARRATIVES WITHIN ONE
UNIVERSITY–SCHOOL–COMMUNITY
"SERVICE LEARNING" PROJECT

Marianne N. Bloch, I-Fang Lee, and Ruth L. Peach

Introduction

In the discursive construction of recent educational reforms, partnerships embody a wide range of differing political, economic, educational, theoretical, and cultural notions. In addition, the word encompasses a variety of reforms within the educational arena, ranging from corporate sponsorships to enhanced family involvement in the affairs of schools (e.g., parent volunteers in classrooms, parent–community–school councils, fees for books or school uniforms, charitable contributions by individuals or businesses). The notion of *partnership* also includes a variety of initiatives between nonprofit community volunteer groups and schooling, including neighborhood center or religious group relationships with schools, partnerships between family literacy councils and schools, and other nonprofit agencies. In the United States, neighborhood agencies and community programs, universities, and schools are examples of these partnerships.

This chapter focuses on one case of a partnership between a nonprofit agency and a school in the United States: a university–school partnership. This partnership was developed to provide additional services—in this case university tutors—to schools in order to support the work done by schools to help children learn and achieve. The partnership represents a renewed emphasis on the sharing of responsibility for education by different members of civil society, a discursive shift associated with Third Way politics in the United States and Great Britain (see Giddens, 1998). It is also part of broader discourses that circulate globally about shared responsibility, alliances, and networked societies (Appadurai, 1996; Rose, 1999).

While university–school partnerships can encompass various reform efforts, the university partnership or *collaboration* that we focus on here is part of a larger partnership between the University of Wisconsin-Madison and several elementary, middle, and high schools that are known as Professional Development Schools or PDSs. The PDSs are a new reform in the United States, and in Madison, in which different programs link the university, and its teacher training programs, with a small group of selected schools in order to strengthen and provide depth to relationships and learning. One of the partnership programs that was initiated in three PDSs in Madison involved the development of a university undergraduate tutoring program that would bring undergraduates from throughout the university into the PDSs at the elementary, middle, and high school levels.

We will describe this particular partnership between universities and schools in a variety of ways in this chapter. We begin with a discussion of the theoretical and historical lens we use to examine partnerships as reforms in education, and then illustrate how those ideas frame our analysis of the tutoring project. The tutoring partnership is one where the university resources (the time and skills of undergraduate student tutors, seminar leaders, and tutoring supervisors) are used as a way to *supplement* and *share* the responsibilities of the school for teaching *all* of their students, and where the school resources (administrators, classroom teachers, diverse K-12 students) are used as a way to *supplement* and *share* the responsibilities of the university for undergraduate education. Some specific university responsibilities for teacher education include helping undergraduates learn about schools, diverse communities, diverse learners, and offering them the opportunity of doing service learning in local communities (see service learning goals, University of Wisconsin Manual on Service Learning, Morgridge Center, 2000).

While partnerships can be seen as ways of bringing two (or more) groups together to collaborate on shared goals and activities, one of the most frequent images of partnership programs seems to fall within a structural arrangement of institutions (e.g., school–family or university–school) to form a workable, functioning organization that helps to maintain and/or to (re)-produce society in the form of good, participating, problem-solving, *productive* citizens. Often, this type of defined look to the future is linked with notions of linear progress that are embedded in ideas of reform. There is no notion of conflict between members of partnerships in many of the new reforms related to educational partnerships to save the young citizen, and nation, nor a questioning of inequities or exclusions in the ways in which organizational relationships and interactions, including schools might help to (re)-produce structural inequalities related to race, class, or gender. This imaginary of partnership, while organized around reforms that may be aimed at improving equality of opportunity within a society, leaves out other critical theoretical framings, and different ways of questioning the material effects of what *appear* as beneficial reforms for families, children, and the nation (see, e.g., Bloch and Tabachnick, 1994; Popkewitz and Fendler, 1999).

Rather than accepting reforms as likely to be beneficial, a conflictual and critical cultural reproduction model examines educational reforms in terms of the possible inequalities in power relations between different partners in a university–school–community partnership and ways in which the state maintains a hegemonic control of schooling, as well as access to different types of schools, and different outcomes. Reforms such as educational partnerships would be analyzed in relation to the reproduction and production of class, race, and gender, and linguistic inequities. Using this equity framework, in a university–school–community partnership, the goals and strategies of one or another partner would be examined in terms of inequalities of power, as well as the ways in which the partnership goals and strategies serve the needs of the more powerful partner(s) over other(s), culturally reproducing a state stratified by class, race, and gender. Partnerships, where the power and knowledge that is legitimate appear to be owned or known by some rather than others, therefore, can appear as a reform for change, while actually changing little, or while enhancing unequal relations (Borman and Greenman, 1994). Many home–school–community partnerships in which parents are to have equal roles with teachers, staff, or university partners embody inequitable processes and outcomes, despite the intentions of reforms (Bloch and Tabachnick, 1994).

Power/Knowledge and A Cultural System of Reasoning Framework

Our service learning project, the SHAPE Tutoring Project, was a university–school partnership that was also born of somewhat unequal understandings, relationships, and relevance/meaning for the university and school partners and actors. As in a conflict model, briefly discussed earlier, the reform of tutoring as an aid to schools, supplementing teachers' work with students, embodied a partnership that had different ways of defining and engaging in relationships, different understandings of problems to be solved, and different motivations for engagement. It was a partnership of unequals, with school staff holding the reins of power, allowing us access into schools and classrooms to talk with students, or with teachers, and defining the ways our students might support, tutor, or mentor individual students, identified by the school staff by agreement. However, the university held power in that it could provide student helpers for teachers, who suggested they were sorely in need of aid in classrooms with large numbers of students with very diverse needs. In addition, the university could be conceived as powerful in that to be partnered with the university in an educational reform offered the possibility of expertise, or, what Foucault (1980) calls *authoritative knowledge* that could be used to help the schools and teachers *save* low-income, limited English-speaking, and non-white students who often were struggling to live up to the standards of schooling present in current American schools and society.

While the inequities in partnership reforms are important in our thinking, they embody a notion of *sovereign power* in which some hold power, while others don't (Foucault, 1979, 1980). Instead we focus on *social inclusion/exclusion* as a part of a system of cultural ways of reasoning about reform as a partnership. We think of reform as an effort between different members of civil society to fabricate a better democratic society, but examine the notions of reasoning embedded not only in reform as progress, but as ways of reasoning about society, community, and a democracy that is inclusive, more generally. By examining power in its relationship with knowledge, we can interrogate how we understand the notions of reform as change or embodying progress. By thinking of reasoning as related to power, we can interrogate taken-for-granted binaries such as inclusion/exclusion or public/private as systems of thought, where power/knowledge relations discipline and govern reason and conduct. This notion of power then draws on what Foucault has termed *disciplinary and bio-power* as the way we reason becomes internalized and inscribed on our bodies, desires, and mind (Foucault, 1979; Rose, 1999).

Partnerships as a Fabrication of a Third Space between
Public and Private

As Nicholas Rose (1999) suggests by examining community in relation to civil society, the notion of partnership exists conceptually within a fabricated third space, (*not* related to Bhabha's, 1994 notion as we use it in this section). The third space appears as a separate sphere, different from the state (or public) sphere and the family (or individual and private) sphere. Civil society, as a separate entity between the state and family, theoretically and organizationally, is a set of actors and institutions that, through free association, can act for the good of the nation and individual, without invoking a central state/government. Conceptually, it also introduces the idea of volunteerism (within civil society) and goodwill as a form of self-governance in the minds of individuals who become active partners/citizens linking the public sector with the private sector. In a remaking of Anthony Giddens's (1998) ideas for a Third Way, members of civil society are required to collaborate and participate to form new ways to care for and educate citizens when the contract between the social sphere of the public or state and the private or family/individual falls apart in the construction of well-being and the formation of well-educated citizens. Educational partnerships fall into this space.

The notion of partnership typically involves a *sovereign* notion of power and knowledge relations, where power can be owned by the state, and distributed to members of civil society or the family; notions of decentralization, and empowerment follow this reasoning. However, sovereign power limits our understanding of the multiple layers of partnerships, as well as the relationships between knowledge/power that assume a social administering by the state from a distance even when it appears that power has been localized or given to others. Drawing on Foucault's notions of governmentality (Foucault, 1991) and power/knowledge relations (Foucault, 1980) as they construct cultural systems of reasoning, Rose (1999) examines the reasoning of partnership as part of the circulation of discourses that currently encourage active citizenship, volunteerism, community service, responsibilization, and collaboration. His discussion of the decline of the social state, and the rise of new networks of associations within what is termed a third space of civil society forces us to examine the discursive framing of our own reasoning and conduct. Drawing on a history of the social administration of freedom (see Bloch and Popkewitz, 2000; Popkewitz, 1998; Popkewitz and Bloch, 2001; Rose, 1999), we can acknowledge the ways in which experience and conduct, even in private, is contingently and discursively produced (see, also, Scott, 1991); this, then, breaks down binaries between public/private that are integral to the concepts of a separate third space of civil society as these are expressed in the Third Way (Giddens, 1998) and other recent neoliberal reforms. It forces us to reexamine the ways in which different technologies of state administration culturally construct and constrain the reasoning that frames the bodily mentalities that fabricate a desire to partner, volunteer, do service to others. It forces a reexamination of our private or personal sense of self-esteem and responsibility for others that appear as natural, necessary and good, when we try to save or provide service to others. It also forces us to reexamine the desire to be individually responsible, to be active, participating members of a third space in communities, part of different networks for reform to save others for the nation.

The Networked, Flexible, and Responsible World

In an increasingly globalized, flexible, and networked world, therefore, the discourse of an imagined safe Third Way—a civil society between the public state and the private family—and of partnerships to save ourselves, nation, and the world is seen as part of new governing patterns of reasoning; these new systems of reasoning or knowledge are seen as related to power. From this vantage point, it is the systems of discursively organized reasoning that construct ideas, identities, and actions, including beliefs about ourselves and others; the same systems of reasoning, at the same time, appear to be inclusive (saving the student from being at risk for failure), while excluding by naming those considered abnormal or different (Popkewitz and Lindblad, 2000).

The internalization of the desire to become a member of a community shifts a group's or individual's identity to that of a responsible and productive citizen who values and plays a role of a partner to facilitate, through local community action and alliances/partnerships, a better societal outcome/vision, a well-intentioned goal. The notion to save future citizens through partnering with them as tutors in schools, however, is at the same time a notion of cultural redemption or salvation that is apparently the mark of responsible, desirable, and good citizens for diverse democracies. In addition, the partnerships of the third space or civil society today are often ones in which the future can be saved for society by educating the *abnormal* citizen (young or old) to make him/her *normal*.

However, the discourses of salvation, responsibility, and service by members of civil society are not new discourses, even though their meaning today is related to the particularities of this moment, this space and place. Before turning to an examination of the SHAPE Tutoring program, as one case of a partnership program that embodies the reasoning of salvation, service, and responsibility, we briefly examine discourses of salvation, responsibility, and service from the late nineteenth and early twentieth century. We use a cultural historical approach to examine reasoning of the past to help us open up the reasoning of the present.

A Cultural History of Socially Administering Freedom

The *discourses of salvation, responsibility, and service*, were important components of the social administration of morality and normality in the late nineteenth century in the United States and in many nations where charitable interventions, religious salvation, and missionary redemption efforts, as well as other forms of education, were used to name and tame the uncivilized and abnormal souls of those considered different. In the United States, during the nineteenth century, it became fashionable to use private financial and other resources to normalize and save those who were considered different in order to save the nation through assimilation and "Americanization" of those who were different. When faced with massive new immigration by Southern Blacks to Northern urban areas, and from the massive immigration into the United States from Ireland and Eastern and Central Europe, the goals of male and female philanthropists were to intervene into family life and fabricate homogenous citizens who could participate actively in, *but in conformity with*, many standards set by the seventeenth and eighteenth century immigrants to the United States (see, e.g., Franklin, 1986). Charitable kindergartens, settlement house nursery schools, parent education classes, and an increase of public primary schools were some of the pedagogical ways that governing was accomplished to teach what is known as standard English, reading, writing, and moral conduct.

Deleuze (1979/1997) and Rose (1999) suggest that by the end of the nineteenth century and the beginning of the twentieth century a social state replaced the individual and group philanthropic or charitable efforts to save others in many Western European nations, as well as in the United States (also see Foucault, 1991; Popkewitz and Bloch, 2001). The social state was characterized by a variety of social laws, policies, institutions, experts, as well as new disciplines of expertise related to notions of science, reason, progress, and truth. There were new ways to assess and categorize groups and individuals—constructing sameness (the poor) as well as difference, normality as well as abnormality. While the strategies of governing were somewhat different in different nations (see, e.g., Wagner, 1994), the move toward a social state that could use a variety of technologies to socially administer freedom and govern individuals for the welfare of the imagined nation was associated with the rise of modernity and modern nation-states.

As the liberal philosophies of the social state emerged in the United States (see Popkewitz and Bloch, 2001), the new strategies of governing encompassed a wide variety of technologies that, together, constructed what governing and care for citizens and nation should be. These included the growth of the modern public school, new laws, policies, and social welfare institutions, gendered, classed, and racialized discourses that framed and constrained family, male/female, and child identity, conduct and experience (Gordon, 1994; Mink, 1995; Skocpol, 1992). In addition, in the early twentieth century, technologies of governing involved the development of new testing and observational techniques to assess normality and development (Danziger, 1990). Governments as well as individuals were discursively directed to save children (and the new democratic nation) from being morally, socially, and academically different. Day nurseries, kindergartens, public schools, settlement houses, and supervised playgrounds were to keep children off the streets and out of factories, to teach a common language, customs, morality, and conduct (for more detail, see Popkewitz, 1998; Bloch and Popkewitz, 2000; Popkewitz and Bloch, 2001). The strategies that were used focused on discursive reasoning that was internalized by teachers and teacher educators, taught to parents, and incorporated expert scientific knowledge to establish appropriate behavior, beliefs, and conduct. As the categorization of normality for children and families was made, the construction of abnormal children and families in need of intervention to be saved and to become normal was constructed.

Volunteerism and Service
As suggested earlier, calls for philanthropy and a patriotic duty to give to one's country were part of the discursive language and practices of the late nineteenth century as private philanthropic and religious efforts were brought to bear to save children and families, and to mold them into images of the good democratic (and homogenized) child/family.

With somewhat different discursive layering, the call for service and volunteers also appeared in discourses of the 1930s in the United States with the depression-era call for a new volunteerism. Franklin Delano Roosevelt's Citizen Conservation Corps (the CCC) was likened to a civilian army of volunteers that could build, restore, and add to the nation through responsible service (as opposed to irresponsible unemployment, poverty, or socialist organizing activities that were considered radical and dangerous). During World War II, Rosie the Riveter presented a textual imaginary that called for women's patriotic volunteer service as a domestic army to work in the American war industry; this image of employment was also a service to save the nation. Women were discursively urged out of other forms of employment, including domestic care for their own and others' children, and into wartime industries. In the post–World War II period, the discourse of service was reversed, with women's patriotic duty becoming service to family, children, and home. Women were urged to volunteer to give up jobs, and return to what was constructed as their natural responsibilities of caring for children at home (see Gordon, 1994; Mink, 1995; Skocpol, 1992).

In the post–World War II period, John F. Kennedy's and Lyndon Johnson's 1960's initiatives—including the Peace Corps, Vista Volunteers, and an Urban Teaching Corps—renewed the call for individual service to nation, and local community control to recreate local forms of social action and democracy. The call to serve and save the country at home, as well as abroad, embedded conceptions of secular service, and notions of individual responsibility and enterprise into the narratives of citizenship and nation. The calls to action were also embedded in the American civil rights movement where marginalized groups (poor/non-white) formed uneasy alliances with white, liberal, northerners (often students) in which the northerners imagined their intellectual and political responsibility to be one of salvation through empowerment, liberation, political resistance as well as social activism.

Social Administration of Normality through Salvation, Service, and Volunteerism in the Late Twentieth and Twenty-First Century
While the child (and family) to be "saved" in the early twentieth century was the different child who was to be homogenized into a universalized American citizen, in the twenty-first century the technologies of governing focus more attention on a democratic pluralism, and a democracy of associated individuals and communities. While we test toward higher standards for everybody, and try to normalize children and teachers so that their performances are in line with standards of proficiency or non-proficiency (another way of designating normal/abnormal), we continue to intervene into the lives of families and children—from the prenatal testing of genetic abnormality to programs for at-risk teenagers to save children—and to fabricate more normal citizens. Using scientific research, statistical and populational reasoning (Hacking, 1991) that emerged in the early twentieth century, we now turn toward a medicalized language about the normal/abnormal child and about how to fix and normalize bodily differences and desires. Through educational

literature, teacher and parent education, and the popular media, multiple discursive layers intertwine to construct the normal child, family, and communities of today who are flexible, able to solve problems, independent, achieving, and yet conforming and responsible. The same reasoning of normality constructs the abnormal, the different child, family or community who is at risk for failure to achieve, or to be independent, responsible, and conform to the new imaginaries of the good democratic citizen of today and of tomorrow.

The saving of future citizens, the guarding of normality, and difference, are technologies of governing reasoning, conduct, and action. The new citizens to save and be saved in the twenty-first century are those who need to learn to be independent, successful (in, out of and after school), responsible, flexible, and cooperative within a diverse democratic society. The good future citizen, embodied in the discussion in this chapter as *both* schoolchildren, and as undergraduates, are to *use their resources* to problem-solve about society's salvation, and, once again, to *serve* the nation as responsible partners, volunteers, and individuals. They are members of constructed third spaces, where flexible partners, members of an imagined homogeneous community, serve to support and save members who are perceived as different for the future of an imagined nation (Bhabha, 1994; Rose, 1999).

Recent reforms, heralding a return to the idea of volunteerism and individual/community action, include a variety of programs, some of which appear similar to those of the past. These include: the America Reads program that brings tutors into work with K-3 grade-level children; Americorps (also calling some of its volunteers VISTA Volunteers, see figure 9.1) that uses recent university graduates to work in schools and other community settings (e.g., helping to recruit and work with parents and other volunteers, work on environmental projects); Retired Senior Volunteer Programs (R.S.V.P.), One Hundred Black Men, a nonprofit group with chapters across the nation; and renewed calls for charitable donations bringing the private members of the third space of civil society, as well as families in to contribute by becoming partners in saving others. A recent cover picture in a local magazine illustrates the discursively implied link between a call to duty, individual power, and responsibility to "volunteer" for one's country. The images of superhero and volunteer are together; volunteers are portrayed as part of a heroic tale, part of the discourse of rugged individualism and enterprise, all of which will, theoretically, save the child and nation as one's self.

Despite a discourse of independence, autonomy, and individuality that is within the notion of the new volunteerism, the discourses of the

PARTNERING TO SERVE / 247

Figure 9.1 Volunteers for America (illustration by Carolyn Fath, copyright, *Ithsmus*, 2002)

early twenty-first century also emphasize partnership, collaboration, and the need for co-construction of ideas. This discursive shift maintains the emphasis on individuality, self-care, and responsibility, while becoming flexible about how failures are described and dealt with in new reforms. Thus, site-based management schemes, parent–teacher councils, and partnerships between nonprofit agencies or other groups in civil society and schools to save children and their families by making them normal embed discourses of normality through a network of relationships, languages, texts, policies, and reform actions.

On January 10, 2002, a "Mentoring A Child" postal stamp was issued by the United States Postal Service. In the newsletter about this stamp's release, the language surrounding the notion of mentor/tutor and partnership suggested the stamp was issued as a way of "raising public awareness of mentoring, helping, and volunteering to make a positive difference in the lives of young people" (see, http://www.usps.com/news/ 2002/philatelic/sr02_007.htm, January 10, 2002). This image on the stamp is of an adult/educated citizen who is willing to volunteer and to join in partnership to mentor a child within different communities. The text from the website cited earlier suggests that mentoring and volunteering are good for all citizens in helping with the preparation of citizens for the next generation.

> With the Mentoring A Child stamp, the Postal Service continues its tradition of highlighting social issues and honors the volunteers and organizations that sponsor or participate in mentoring programs through organizations that include Communities In Schools, Girl Scouts of the USA, 100 Black Men of America, Everybody Wins!, Big Brothers Big Sisters of America and Boys & Girls Clubs of America, as well as schools, religious communities and corporations.
> Volunteers share constructive and fun activities with young people after school and on weekends, and encourage others to also volunteer. Many mentoring programs are tailored for disadvantaged youth, but all young people can benefit from relationships with committed adults outside their families.

The illustration, of a man and boy (see illustration on website) illuminates how cultural images (and the text on the stamp and website that accompanies the illustration) can suggest identity and conduct of "good" citizens. The stamp also translates the idea of mentorship into a discursive portrait of service and salvation; it requires a construction of who the mentor or model should be, and who is in need of intervention through tutoring, or mentoring/modeling. ("Many mentoring programs are tailored for disadvantaged youth, but all youth can benefit from relationships with committed adults outside their families.") In addition, it requires a construction of who can partner with the school to save the school, nation, and its children. In the new communities of interest, teachers are no longer entirely responsible for problems that they attribute to causes beyond the school in children's home or community life. Nor is the society or the state constructed as responsible for providing for the basic welfare of its citizens, as this, too, in current partnership discourses, appears as a shared responsibility between responsible, independent, and autonomous individual and group actors within civil society.

Current governing strategies embody the broader community as a network of relationships, and in the way knowledge is embedded in a variety of ways as ways of reasoning about action, selves, and others. This *community* governs itself, each other, as well as our own subjectivities. As children come to school from different backgrounds—some more impoverished than others, some more constrained by less than living wages, some more burdened (and made rich) by less knowledge of English/more knowledge of other languages in an English immersion-oriented society, the school and the community are encouraged, discursively, to share in the problem of saving these children and their families. It is from this vantage point of educational reforms as partnerships that our university–school tutoring partnership emerged.

The ways that we reason about ourselves and others, who we construct as normal/abnormal, deficient, or as successful, are critical to our examination of the SHAPE Tutoring program, a project we have worked with for five years at the University of Wisconsin-Madison. Using Foucault's theoretical framing as a background for our analysis, we question taken-for-granted notions. Drawing from Derrida's (1981) ideas of deconstruction, the analysis examines ways of reasoning and the language used to describe the project and the pedagogy associated with it, the researchers, teachers, tutors, and tutees, and the perceived partnerships between each of the project participants to engage in a partnership to do responsible community service, and to help to save children through enhancing their learning/success in schools.

In the next section, we illustrate these ideas by describing and analyzing the reasoning inscribed within the SHAPE Tutoring program. We analyze the stated description of SHAPE and its goals as a university service learning project and as a partnership between the university and the community, the policies related to selection of tutors and tutees, the seminar used for tutor training, and excerpts from tutors' and tutees' statements and actions as these were recorded by tutors writing about their experiences. In an effort to examine the notion of normality and success, we also examine statements made by researchers, tutors, and teachers related to perceptions of tutees' success in school, as well as the tutors as models of success.

SHAPE: A University–School Service Learning Partnership to Tutor Future Citizens

We use the SHAPE Tutoring Project as one concrete way to illustrate the points we made earlier related to discourses of a third space for partnerships, salvation narratives, and the discourses of responsibilization and service. SHAPE is a current university-based service learning program and university–community partnership that has brought undergraduates from the University of Wisconsin-Madison to six local public schools that are defined demographically as diverse in terms of the economic, ethnic, and linguistic background of students. We, the authors, are directors and staff of the project.

The name SHAPE was developed in the first semester of the program by the undergraduate tutors themselves; it initially stood for "Students Helping in the Advancement of Public Education," and, after two years, was changed to "Students Helping in Achievement in Public Education." The program started in 1997 in response to a student activist group's call for the university to increase efforts to diversify the campus by increasing undergraduate enrollment of more students of color. One point in the undergraduate student activist group's action plan called for the development of a program that would link undergraduates as tutors/mentors with students of color in diverse public elementary, middle, and high schools in Madison, Wisconsin in order to help more children from low-income, and African-, Latin-, and Southeast Asian American backgrounds obtain better grades in public schools.

The Discursive Language of Service and Partnership
SHAPE is designated as a service-learning opportunity for University of Wisconsin-Madison undergraduates. While not a requirement for any major on campus, participation in service learning courses is currently encouraged as valuable, responsible service/learning across the university. For SHAPE, students sign up for 1–2 credits of coursework, take a two-hour weekly seminar on campus, and tutor in elementary, middle, or high schools with designated tutees two times per week (2 hours/credit). The seminar requires readings, journals, and a portfolio describing tutoring experiences, discussion, and attendance.

In the *University of Wisconsin-Madison Manual on Service Learning* (2002), three necessary criteria for academic service learning are stressed: (1) *"Relevant and Meaningful Service* With the Community: service provided in the community must be both relevant and meaningful to all stakeholders. There is *purposeful collaboration between the University and the community. And the community plays an active role in defining what the students' service activities will be*; (2) Enhanced Academic Learning: the addition of *relevant and meaningful service with the community must not only serve the community but also enhance student academic learning* in the course. The service and academic goals must inform and transform one another; and (3) *Purposeful Civic Learning*: the addition of relevant and meaningful service in and to the community not only serves the community and enhances student academic learning in the course, but *also prepares students for active civic participation in a diverse democratic society.*" (Abstracted from the Michigan Journal of Community Service-Learning, summer 2001, and cited in the Wisconsin manual mentioned earlier; Italics added for emphasis.)

SHAPE was designated as a desirable course for students, both by advisors interested in some students having experience doing service in schools or in the community (outside campus), and by students and administrators as a way to promote the idea of and contribution of service to schools and children as part of the broader community and nation. Undergraduates who chose to be tutors often chose to tutor children in schools to do good (and *become* good) through service that also embodied a salvation narrative. Our work with students was in part to deconstruct that narrative; however, we also reinscribed it simply through discourses of serving others perceived or constructed as different.

The Discursive Practices of Partnership
The service learning manual mentioned here prioritized relevance and meaning to all stakeholders, the importance of purposeful collaboration between the university and community, and the necessity that the community plays an active role in defining what the service activities will be.

SHAPE was a partnership that embodied many of the possibilities as well as problems of the discourse of partnership and individual service, when different institutions and actors with different interests, and responsibilities form an alliance. School administrators and teachers were eager to have help in classrooms that were very diverse in terms of student ethnicity, income, gender, ability, first language, and so on. They agreed to the SHAPE program readily. However, they were less able to take time to work with SHAPE administrators and staff or even with the undergraduates who came to tutor students in classrooms, and were also less interested in one-to-one tutoring (the SHAPE assigned mandate from the chancellor's office) than in having a more general aide with all children in their classes. The SHAPE administration and staff, as well as the university tutors, shared one purpose—learning about diverse groups of children, and learning about schools and teaching for the undergraduates, as well as giving individual supplementary assistance through tutoring or mentoring to children who might benefit from extra individualized attention, *and* who could not afford the price of private tutors engaged for children by many middle- and upper-class families in the United States. Children and parents were stakeholders, too. The children were assigned to be tutored only with their own and their parents' permission, with knowledge that tutoring would be done by university students, and for free.

While time to elaborate shared objectives and effective strategies by all partners, including those tutored, their parents/guardians, and teachers was never easily available, periodic meetings and written communications between the majority of partners or stakeholders occurred with some frequency. Each communication from the university partners reclarified goals and expectations but generally there was little feedback or shared discussion over goals from school partners.

Despite these communications, the undergraduate tutors felt they needed more information from teachers to do an effective job working with students they were tutoring. Teachers consistently expressed appreciation for the tutors' work, and said it made a difference for students who were tutored and for themselves by having extra help in their classroom; however, they said they had little time for extra explanations or directions, and, despite calls for meetings, teachers didn't completely understand what was expected of them, despite the project team's efforts to communicate (Bloch, 2000).

Finally, while the elementary, middle, and high school students were asked whether they would like a tutor, it was after parents had been contacted, and after they had been designated by teachers as students who could benefit from having a university tutor with whom to work, that they agreed to be tutored. As key stakeholders in the partnership, this often left the tutees wondering why they had been selected for such an arrangement (e.g., why do I need a tutor? Am I different from the kids who don't have tutors?), and although infrequent, sometimes resisting the construction of "in need of being tutored" by refusing the assignment of a tutor, or by not paying attention to or avoiding UW undergraduate tutors when they came to tutor. Nonetheless, over 5 years of the program, there were many more tutees who wanted a tutor, than those who didn't; elementary students acted as though they were delighted to have a university tutor work with them, while middle and high school students' reactions varied from pleasure at having a university tutor/mentor to talk to, play sports with, and work on homework with, to active resistance or refusal.

Given differences in race, class, gender, language, and schooling experiences that were common in the tutor–tutee pairings, as 90 percent of the UW tutors were Euro-American, and 75 percent of the tutees were students of color, often with English as a second language, increasing the degree to which tutees participated in program planning might have helped with tutee selection and the development of more effective, and fewer frustrating relationships. However, in addition, the labeling of young students as being "in need" of intervention by a tutor, involved the construction of a difference and abnormality that the tutees understood.

Discourses of Dependency and Difference: Power/Knowledge Relations
While other common difficulties of partnerships arise when different stakeholders have different agendas, experiences, and power (a sovereign notion of power), the fact that student tutees were asked last if they wanted a tutor was related to the construction of children as young, in need of guidance, less capable, and dependent upon the wisdom of elders. It relates to the cultural construction of childhood as a time when children need care and do not need to be asked, and in this way, is an effect of power/knowledge relations related to age and notions of dependency.

Similarly, when teachers had little time to talk to their university tutors or to the SHAPE staff who were supervising the undergraduates, it was related to differential power, interest, and time, as well as the construction of differential expertise and knowledge as the relationships between university knowledge as theoretical/abstract and expert was contrasted to school knowledge as atheoretical/concrete and applied and done by practitioners (who, in relation, could be constructed as nonexpert). These different constructions of partners, developed as technologies of power/knowledge during the twentieth century, are cultural ways of reasoning that appear natural, but clearly involve notions of normality, authority, and difference that affected the program (see Foucault, 1980, on power/knowledge and truth).

Saving the Citizen and Nation through Service to School, Community, and Others: Tutors who are Successfully Tutoring the Child with Potential

Discourses of Success, Normality, and Abnormality
Here we illustrate the cultural reasoning systems used by different partners in the ways in which undergraduate tutors and public school tutees were selected to be in the SHAPE project. We give further detail about how these partnered tutors and tutees perceived themselves, and in relation to the others with whom they were to have a relationship. We also describe the ways in which undergraduates began to learn, in our seminars, to participate in schools to help them learn responsibility to self and other.

The stories were constructed from an amalgamation of different experiences: discussions in seminars, observations of tutors–tutees in classrooms and homework centers, journals written by university tutors, end-of-semester summary portfolios, as well as teacher reports in interviews and grades given to tutees. We have excerpted some of the discussions and observations from several pairs of tutors over this period to illustrate key points about the partnership and the reasoning embedded in all of the partners' actions, conducts, and belief systems. The first section here illustrates reasoning used in the *selection processes* that were used. The second section focuses on the ways in which the undergraduates' *service learning seminar*, an integral part of the service learning course, combined with their community service of tutoring diverse students in public schools. We examine constructions of service to the nation, responsibility/irresponsibility, normality/abnormality or difference, success in school/lack of success, and potential/lack of potential to achieve in school. Finally, we examine the way tutor–tutee relationships, and our own teaching reinforced and reinscribed these narratives in the program.

Selection Processes: The Identification of Tutors' Identities and Subjective Ways of Reasoning
Perhaps because the vast majority of undergraduates at the University of Wisconsin-Madison are from Euro-American backgrounds (90 percent), the majority of SHAPE tutors/mentors have been Euro-American students, too, with 10 percent of the undergraduate tutors African-, Latino-, or Asian-American. A large proportion of undergraduate students speak at least one language beyond English, most frequently Spanish, with approximately equal numbers of male and female tutors going into schools from the university. The SHAPE tutors have come from different majors, including Engineering, Math, Sciences, Agriculture, Social Welfare, Business, History, English, Psychology, and Architecture, but with the majority interested in teaching as a possible future major or profession. They are recruited from all over the university campus in order to "model" different majors, and to bring different kinds of expertise into schools. The majority of the tutors have not had much experience working as tutors, nor with younger students from linguistic or ethnic/cultural backgrounds different from their own. Most of the UW Madison tutors are also good students, and know how to play the student "game," having been admitted into a highly competitive university that requires good grades, national standardized test scores, and the knowledge required to behave as a good student (doing homework, decent attendance, a Bourdiean-like *habitus* that appears helpful to school success). While these students appear to be ideal tutors or mentors for public school students perceived as having potential, they

often turn to tutoring with the idea that poor students are in need of help and can be saved by teaching them how to act and behave as they have been taught to act and behave in homes and schools.

Tutors are assigned to classroom teachers, and eventually to students by administrative staff and/or teacher self-selection. Within classrooms or within schools, administrators and teachers were asked by the university SHAPE administrators to select students whom they perceive to be behind but, with help from a tutor/mentor, might have the potential to achieve better in school. This definition of who should be tutored was typically worked out by a meeting of university and school administrators, but often without teacher, tutor, or tutee presence. However, teachers were the primary people who determined who should be tutored once tutors were assigned to a classroom.

The Selection of Tutees: Having the Potential to Succeed or Not
At first there were no other defining criteria to the term "potential" as we all hesitated to designate potential based upon grade point average, or other artificial markers of competence, intelligence, creativity, or likelihood to succeed in the current modern public school, where attendance, homework, grades in tests, and behavioral conduct or habitus, and family background (cultural capital) seem heavily related to the construction of at risk or with potential, or even success (Bloch et al., 1994; Bourdieu, 1986; Bourdieu and Passeron, 1977; Lareau, 1989; Swadener and Lubeck, 1995; Tabachnick and Bloch, 1995). This, however, became a problem as teachers and administrators defined the term in many different ways, including resistance from one elementary school to the attempt at making a discursive shift from *at risk* to *with potential*, by suggesting that all children had potential; how could they choose who should be tutored, they wondered? In other schools and classrooms, teachers selected their most troublesome students, and defined them to us as students *with potential* in order to obtain a university tutor for the student. Students who were selected as tutees were from low-income families (based on criteria of free/reduced lunch) and/or limited English proficiency students, and/or low-income students of color. The many tutees SHAPE tutors worked with over the full five years of the program (1997–2002) discussed here have been in grades Kindergarten through tenth grade and are from diverse backgrounds—typically low-income, Latino, Hmong, African American, American Indian, and Euro-American.

Although we asked teachers and administrators to select tutees with potential for achievement who might benefit from having a university tutor/mentor, we found that the majority of tutees at all levels (elementary, middle, and high school) assigned by teachers and school administrators were students who were constructed as at risk for problems in literacy or math, who were having trouble turning in homework, school attendance, or grades (despite some indicators of academic competence or social leadership), or were students who were not conforming to school or classroom rules and conduct, but appeared to have vaguely defined potential to do better. With our directions (organized with school administrators) to identify children on the basis of different types of potential to do better with the help of a university tutor, the university staff accepted diverse notions of potential as tutees were selected and named as tutees in the program. In an addition to the program, during the second year, selected tutees at middle- and high-school-levels were also labeled *SHAPE Scholars*, and the tutees, their teachers, parents, and undergraduate tutors were told these were students identified by school staff and by our project as those with potential to achieve better in school, and with potential to move into higher education.

The university staff reclassification, in collaboration with school administrators, of the children as having potential, while an act to counter the schools' classification and construction of the children as not doing well, and abnormal, was a discursive shift in language that used Foucault's concepts of power/knowledge and power in relation to truth (Foucault, 1980) to reconstruct children's self-identities, and the identities assigned to them by others (teachers, administrators, tutors, and parents/guardians). Our question was: would it work? It was only at a later point that we came to understand this as an intervention to save students identified as different or at risk that embodied our own salvation narratives as part of the intervention and reform.

The Service Learning Seminar as a Narrative of Partnership and Volunteerism

The service learning course gave credit and required training, and critical reflective discussion of readings, observations, and experiences in schools and classrooms with the students who were tutored or mentored. Despite this broad goal, the program strategies were normative and instrumental in that we began each semester's seminar for new tutors with training on tutoring in general, how to enter into schools and classrooms, beginning relationships with tutees, knowledge about literacy and math, how tutors might work with students to become more successful—how to help them acquire habits of the *good* student by completing homework on time, becoming better organized, or knowing when to speak up and ask questions, and when not to.

Along with these normative and very traditional tutoring program goals, SHAPE tutors were also pushed to reflect critically on their readings and work with SHAPE tutees, and within school and community settings. Readings focused on issues of race, class, language diversity, and culturally relevant teaching, including the need for advocacy for students within classrooms, and teaching the codes of power related to getting along in school (Delpit, 1995). Course readings touch on linguistic and cultural diversity (e.g., Tse, 2001; Lee, 1996; Ladson-Billings, 1994, Tabachnick and Bloch, 1995), and focus on the strengths of students, attempting to deconstruct narratives of salvation and deficiency and abnormality as we saw these appearing in seminar discussions, weekly student journals, or more informal discussions with us or that we observed.

Despite our attempts to deconstruct and reconstruct discourses used to describe and inscribe student identities by talking about strengths of students or about students with potential, our tutors, working with tutees identified as at risk or with potential, were constantly enmeshed in a reinforcement of salvation narratives, and discourses that alluded to the need to save the abnormal child, who has potential for achievement, for success in schools.

While problematized from the beginning, the SHAPE program, by using university undergraduates, in many ways incorporated the idea that someone with knowledge about "doing school" and who was successful in academics and school-type behavior, could or should teach these skills, reasonings, and conduct to another; the program also assumed such a student could or should be a good tutor/mentor despite many differences in experiences, attitudes, and skills in our undergraduate tutors over the five years of the program. Thus, the idea that all undergraduates—the universal good student—had skills, conducts, and knowledge that embodied successful student behavior was part of our program reasoning. In addition, we initially left unquestioned the idea of needing to identify the tutee with potential to succeed. Not knowing what success meant, except in the traditional terms of school report cards, grades, attendance, promotion records, and good conduct reports, we asked administrators and teachers to nominate tutees with potential, leaving the term vague. However, the need to identify those to be tutored or mentored—or those in need of help through tutoring/mentoring activities—became an ongoing source of conflict as we understood the ways in which we reinforced the linking of a student being a child of color with abnormality or at risk simply by using the reverse reasoning of with potential. While we tried to be inclusive in our reasoning about diverse strengths and potentialities that could be successful, we also reinforced the reasoning of normality/abnormality, which was narrow, bounded, and closed the identities of children who, using this reasoning, *could/should* be saved from these differences.

One of the problematics embedded in the partnership became the way we all shared similar discourses and practices, thereby limiting possibilities for rethinking the unboundedness and multiplicities of identity, diversity within cultures/ways of living/behaving. When the discourses were critically examined, we found ourselves still reinforcing notions of sameness/and difference, and cultural redemption through salvation of others. These ideas of doing service, being responsible, volunteering and being a partner, as well as those of normality/abnormality, and salvation, circulated through our program, despite our intentions to deconstruct and reconstruct them into other ways to envision relationships, educational reform, or the multiplicity and richness of identities.

A Partnership in the Discursive Framing of the Construction of At Potential/At Risk

While UW tutors were told they would tutor students designated as "with potential to achieve at normal or above normal levels of achievement," within classrooms and schools, university tutors' discussions in seminars, journal entries, portfolios, and exit interviews often suggested that they were being assigned students designated as having potential but also who were considered to be at risk for failure. Thus while the community partnership between university and schools was originally designed to encourage university students from all fields to tutor in schools in order to learn about community, diversity, and different potentialities or strengths of students, the SHAPE program appeared to reinforce conceptions of deficiency and discourses of salvation—that the tutors could save the abnormal student, by making him toe the line in school, or *be normal*. A few stories, reconstructing some of the tales told by tutees, tutors, and teachers, as well as our own interpretations of these tales, illustrate these points about the cultural reasoning embedded within students' relationships.

> *Joe: I Am the "Bad Boy."* One of our tutees, a young African-American male 7th grader named Joe, for example, was considered a troublemaker when he spoke out of turn in class. While he had participated actively and with great interest in some of our out-of-school activities with tutors and mentors for over a year and a half, he told us one day that he could

no longer participate in our tutoring program, or the extracurricular activities that he had appeared to enjoy. Why? He told us he was being sent to "bad boys school" (his term; however, in the school counselor's term, this is an alternative school for children with special needs) and had become ineligible for tutoring, or the privilege to participate in extracurricular activities including our tutoring and mentoring activities.

From one perspective, this was an example of "tough love," so commonly touted as necessary in today's (urban) schools (see Popkewitz, 1998, on the notions of "urban schools" here); from another, this young child (only 11 years old) had been categorized and disciplined, and he had internalized, to some extent, others' messages—"he was a bad boy"— according to others' constructions of him, and now his own, and was literally excluded, even from other activities he might have enjoyed and that were open to him.

Mark—A Poor Student or A Student with Potential? Similar to Joe's internalization of himself as a "bad boy", Mark, an 8th grader, claimed/announced himself as slow to his tutors after one of his regular after school tutoring session at a homework club: "*You don't really know me. I am very slow!*" While Mark's report cards crafted him as a student who is not turning in homework and assignments on time, he is being perceived as a student who is capable of doing better in some of his teachers' eyes. Throughout a period of 2 years in the tutoring project, across the different tutors, Mark is constantly being described as a student who is eager to learn and who often initiates a transition from social greetings into working on his homework. Despite Mark's efforts on his homework assignments, the comments on many of his report cards tell a different tale. Comments like: "assignments late or not complete, grade affected by missing assignments, and needs to do homework" have became the most typical remarks in his report cards.

With the comments of Mark's report cards in mind while reading his tutors' reflective journals, which often portrayed Mark as a smart student who always comes to the tutoring session with energy and ready to work, there appeared to be several missing links: (1) between Mark's initiative in asking for help with doing his homework and the teachers' records of his missing homework, and (2) in the partnership involved in shared communication and discursively shared imaginaries of Mark, his university tutor and his teacher.

Keshandra—A Student with Potential. Keshandra has been identified by SHAPE tutors and by school personnel as a bright young woman with potential. She is African-American, the child of a single mother who works at the local university. Her mother knows she is bright and wants her to go to university; Keshandra asks so many questions in some classes (e.g., social studies) that her teachers often cannot answer them. When we first met her, she proclaimed her intention of becoming a lawyer, which was supported in multiple ways by her school, including a pre-employment program in which she shadowed a working lawyer, and a summer internship with a lawyer. Pat, a young white male, was assigned to help Keshandra in math. He was a very thoughtful and a caring young man who had struggled in high school himself, and who saw great possibilities for (working with/saving) Keshandra.

(Pat) "... I didn't, I didn't really get to understand her, I never, I didn't realize just how, how smart she was for awhile. I still, I had, I think I had, I had some preconceived notions, you know, if she needs a tutor she's probably not that bright ... but I think I eventually I figured out that, you know, she just doesn't know how to play the system, like I do. She didn't realize that getting some extra credit points could bump your grade up ... I told her that and she was amazed!"

The Tutees under Different Eyes

Throughout the course of the tutoring project, different members of the partnership were able to see different aspects of the SHAPE program—the tutoring, tutors, teachers, and tutees—as these were constructed through different lens or eyes. Through the grades and teachers' comments on the tutees' report cards, one side of the tutees could be constructed, while through the tutors' reflective journals, essays, as well as the interviews, another side of the tutees could be constructed. In the next story, we illustrate how the discourses that portray identity and conduct shifts, and are unbounded, as well as uncertain, when multiple stories are at play at once.

A Tutors' Story—The Goal: Students Who are Successful
The constructions of normality/abnormality take place in our everyday schooling practices. In our tutoring program, teachers asked tutors to help tutees learn appropriate skills, conduct, and to model good student behavior—as after all, *they* were successful students—*they had been admitted to the University of Wisconsin*, where admission is based on culturally biased grade point average, rank in school, and ACT or SAT scores—true marks of success! The tutors, our UW undergraduate students, were asked to form a partnership with the school, the teachers, and the tutees to develop future "good citizens" who would also succeed in school—as though 2–6 hours per week of tutoring by relatively

untrained tutors could make up for continuous tracking, poverty and its correlates, and two centuries of educational pedagogies and policies that discursively framed people of color as abnormal and deficient.

What we came to know and learn about each individual child as a tutee/student was through his/her teachers' eyes as well as through the tutors' eyes, and our own reconstruction of discursive texts. The understanding of the tutee as well as the tutor and other partners' perspectives in the program was *always* partial, and, as a prism, with many sides to it. By crafting a tutee as a child with potential, we had hoped to be able to open up new frames of reasoning that would allow new identities, new paths for conduct and reasoning. We had hoped to move toward a moment when different participants could be seen as in process, dynamic, and as subjects with multiple sides, positions, and non-constrained identities. In the end, as partners in SHAPE, we found we were unable to break far from a discursively organized, contingent space and place we all shared; the common terms and meanings of the project with undergraduates, school staff and teachers, and even with many of the tutees and their parents were embedded in all of our ways of reasoning.

The shared discursive spaces of *doing service* or *partnerships to save others* that were to make us responsible members of a pluralistic democratic community are the spaces in which most of today's educational reform discourses circulate. However, these discourses limit possibilities, rather than opening up new spaces for reason and action.

It is clear which students have been successful historically, and what effects conformity to "whiteness" or Euro-American cultural norms requires of many students of color and their families (e.g., see Delpit, 1995; Lee, 1996; Tabachnick and Bloch, 1995). We also know of the disciplinary and regulatory patterns in governing that try to normalize, while categorizing, differentiating, and constructing subjectivities of abnormality and difference/deficiency. Therefore, the normal, developed child/student/tutee, parent, teacher, university student/tutor of the twenty-first century is judged on his or her capability to be the responsible, flexible, and autonomous child, parent, teacher, tutor—one who can behave responsibly, morally, and flexibly, bending to new rules and opportunities, be self-problem-solving, as well as self-disciplined, partnering to save self, community, and nation. Within the United States, this child/parent/teacher/tutor, however, is still cast as the imaginary of the essentialized good (white, middle-class) conforming body who should or has achieved well in school(s). While the state governs through mandates for testing, as well as higher standards for all children (e.g., *No Child Left Behind* policies), the flexible school works or is asked to work with a third space—an independent, and flexible set of partners in the community. Families, and children are to meet standards constructed from a distance—constructed for the universal child, the universal citizen of tomorrow, the universal imaginary of the good capable, economically autonomous, and responsible citizen. The perceived freedom and autonomy of identity and conduct remains one that is socially administered to be one of normality and a sameness that fails to recognize (let alone appreciate) difference.

Rethinking Community Partnerships as a Way to Save the Child of Tomorrow: Unpacking the Meanings of becoming a "Tutor," "Mentor," "Tutee"

Tutoring and mentoring programs have had many apparent achievements as we look across their history. Some of our own tutors/mentors have told us they are in our program because of the positive influence of tutor/mentors they had in their own life. However, we need to unpack these words for a different understanding of the meanings that are embedded within the words. For example, to be a tutor means that one ideally *can* tutor (a skill, behavior, attitude), and that one *may* tutor (the tutor is offered the chance to tutor, has an opening to work with a tutee, is directed to tutor). To tutor implies the ability to teach something through example or illustration. To mentor, on the other hand, implies a supportive relationship, a model, and a context of directional care for another, and, again, the acceptance of some relationship that allows for mentoring, modeling normality or success, and the governmentalities that enter those notions of success into the soul of the other. As with the notions of partnership and of doing service, the words imply a need to change, as well as aspects of directionality and reciprocality with and through each other. The words involve governmentalities that construct and constrain the tutors' and the tutees' subjective identities and conduct toward each other. These power relations (sovereign, disciplinary, and bio-power, see Rose, 1999) govern in addition the mentalities of those who work with the tutoring program, as well as those in other educational reforms and partnerships to improve schools, and save children.

Today's reforms in the United States suggest that tutees *at risk* or *with potential* should have a tutor to help them become successful. In fact, this is a way in which richer parents often provide a needed boost for their own children. Because tutors wanted to do service and help others achieve greater success, as currently defined, and some tutees wanted a tutor to work with, someone other than their teachers, the program has continued. While discursive constructions of identity, normality, notions of success, and deficiency, sameness and difference encase many of the older and younger students, strategies to break out of these languages were limited. It was a challenge to find another space, filled with a language of different possibilities.

The Imaginary of Success and the Construction of the Good Citizen through Partnerships

The alchemy of partnership as a salvation narrative cannot be defined, simplified, and held as the model or the solution to educational reforms. In our own experiences as participants and *partners*, we constantly see ourselves working against as well as being immersed in the discourses we have elaborated throughout this chapter.

But, as a service learning project, SHAPE, continues today. How might we shift these ideas? One possibility is to reconceptualize the notions of success, the notion of partnership as a discursive layering, and a continuation of the social administration of identities of who and what the well-educated citizen of tomorrow, or we, should be. We might open up these terms to more critical analysis, thereby also recognizing power and knowledge as this relationship constructs our actions, beliefs, and ways of reason. The notion of partnership is a part of this layered reasoning that needs critical reexamination. The third space of civil society that is to save the future of the imagined nation with a variety of new educational reforms and alliances—also must be a discursive space that is continually interrogated. Rather than a locale—a third space—apart from state or family, must be recognized as a socially administered space of reason, where power and knowledge form the limits as well as open up possibilities for reason/unreason. By deconstructing educational reforms that embody notions of service, volunteerism, responsibility, potential, and partnerships that can save the nation, we can examine them as artifacts of culturally contingent historical places and space. By interrogating their naturalness, we also open up new possibilities for reason and action. It may still be possible to think about how to recognize more of the multiplicities of talents, identities, and possibilities offered by all the youth (school-age as well as undergraduate) with whom we've been working. It may be possible to reconstruct conceptions of success in today's schools and universities.

Such an opening up to complexity of identity and conduct would constitute a reconstruction of the national narrative of assimilation and success, a reconstitution of the national imaginary of the homogeneity of the good citizen, currently based on normative discourses of autonomy, flexibility, independence, responsibility, individuality, and competition. It would require us to recognize the relationships between these narratives and a Western Euro-American conception of modernity, progress, and normality embedded in schools, universities, and society(ies) more generally. Recognizing a third space as a hybrid, complex construction of cultures, identities, and ways of being successful, while difficult, is a critical first step (Bhabha, 1994). Recognizing the multiplicities of identity and strength rather than difference and abnormality would also open new prisms of possibilities, and ways to interrogate what is *reasonable* or true.

References

Appadurai, A. (1996). *Modernity at Large: Cultural Dimensions of Globalization.* Minneapolis: University of Minnesota Press.
Bhabha, H. (1994). *The Location of Culture.* New York: Routledge Press.
Bloch, M.N. (January 2000). *An Evaluation of the SHAPE Tutoring Program.* Unpublished manuscript. School of Education: University of Wisconsin-Madison.
Bloch, M.N. and Popkewitz, T.S. (2000). "Constructing the Child, Parent, and Teacher: Discourses on Development." In L.D. Soto *The Politics of Early Childhood Education* (pp. 7–32). New York: Peter Lang Publishers.
Bloch, M.N. and Tabachnick, B.R. (1994). "Rhetoric or Reality: Improving Parent Involvement in School Reform." In K.M. Borman and N. Greenman (Eds.), *Changing Schools: Recapturing the Past or Inventing the Future?* (pp. 261–296). Albany: State University of New York Press.
Bloch, M. N., Tabachnick, B.R., and Espinosa-Dulanto, M. (1994). "Teacher Perspectives on the Strengths and Achievements of Young Children as these Relate to Ethnicity, Language, Gender, and Class." In B. Mallory and R. New (Eds.), *Diversity and Developmentally Appropriate Practices* (pp. 223–249). New York: Teachers College Press.
Borman, K.M. and Greenman, N. (Eds.) (1994). *Changing Schools: Recapturing the Past or Inventing the Future?* Albany: State University of New York Press.
Bourdieu, P. (1986). "The Forms of Capital." In I.C. Richardson (Ed.), *Handbook of Theory and Research for the Sociology of Education* (pp. 241–258). Westport: Greenwood Press.
Bourdieu, P. and Passeron, J.C. (1977). *Reproduction: In Education, Society, and Culture.* London: Sage.
Danziger, K. (1990). *Constructing the Subject: Historical Origins of Psychological Research.* New York: Cambridge University Press.
Deleuze, G. (1979/1997). "Forward: The Rise of the Social." In J. Donzelot (Ed.), *The Policing of Families* (pp. ix–xvii). Baltimore: Johns Hopkins University Press.

Delpit, L. (1995). *Other Peoples' Children: Cultural Conflict in the Classroom.* New York: The New Press.
Derrida, J. (1981). *Dissemination* (B. Johnson, Trans.). Chicago: University of Chicago Press.
Foucault, M. (1979). *Discipline and Punish: The Birth of the Prison.* New York: Vintage Books.
Foucault, M. (1980). *Power/Knowledge* (C. Gordon, Trans.). New York: Pantheon Press.
Foucault, M. (1991). "Governmentality." In G. Burchell, C. Gordon, and P. Miller (Eds.), *The Foucault Effect: Studies in Governmentality* (pp. 87–104). Chicago: University of Chicago Press.
Franklin, B. (1986). *Building the American Community: The School Curriculum and the Search for Social Control.* London: Falmer Press.
Giddens, A. (1998). *The Third Way: The Renewal of Social Democracy.* Cambridge: Polity Press.
Gordon, L. (1994). *Pitied but not Entitled: Single Mothers and the History of Welfare.* Cambridge: Harvard University Press.
Gordon, and P. Miller (Eds.), *The Foucault Effect: Studies in Governmentality* (pp. 181–195). Chicago: University of Chicago Press.
Hacking, I. (1991). "How Should we do the History of Statistics?" In G. Burchell, C.
Ladson-Billings, G. (1994). *The Dreamkeepers: Successful Teachers of African-American Children.* New York: Jossey Bass Publishing Co.
Lareau, A. (1989). *Home Advantage.* London: Falmer Press.
Lee, S. (1996). *Unraveling the "Model Minority" Stereotype: Listening to Asian-Americans.* New York: Teachers College Press.
Mink, G. (1995). *The Wages of Motherhood: Inequality in the Welfare State: 1917–1942.* Ithaca: Cornell University Press.
Popkewitz, T.S. (1998). *Struggling for the Soul: The Politics of Education and the Construction of the Teacher.* New York: Teachers College Press.
Popkewitz, T.S. (2000). "Globalization/Regionalization, Knowledge, and the Educational Practices: Some Notes on Comparative Strategies for Educational Research." In T.S. Popkewitz (Ed.), *Educational Knowledge: Changing Relationships between the State, Civil Society, and the Educational Community* (pp. 3–30). Albany: State University of New York Press.
Popkewitz, T.S. and Bloch, M.N. (2001). "Administering Freedom: A History of Present Rescuing the Parent to Rescue the Child for Society." In K. Hultqvist and G. Dahlberg (Eds.), *The Changing Child in a Changing World.* London: Routledge.
Popkewitz, T.S. and Fendler, L. (Eds.) (1999). *Critical Theories in Education: Changing Terrains of Knowledge and Politics.* New York: Routledge.
Popkewitz, T.S. and Lindblad, S. (2000). "Educational Governance and Social Inclusion and Exclusion: Some Conceptual Difficulties and Problematics in Policy and Research." *Discourse: Studies in the Cultural Politics of Education,* 21(1), 43–82.
Rose, N. (1999). *Powers of Freedom.* Cambridge: Cambridge University Press.
Scott, J. (1991). "The Evidence of Experience." *Critical Inquiry,* 17, 773–797.

Skocpol, T. (1992). *Protecting Soldiers and Mothers: The Political Origins of Social Policy in the United States*. Cambridge: Belknap Press of Harvard University Press.

Swadener, E.B. and Lubeck, S. (Eds.) (1995) *Children and Families "at Promise": The Social Construction of Risk*. New York: State University of New York Press.

Tabachnick, B.R. and Bloch, M.N. (1995). "Learning In and Out of School: Critical Perspectives on the Theory of Cultural Compatibility." In B. Swadener and S. Lubeck (Eds.), *Children and Families "At Promise": The Social Construction of Risk* (pp. 189–209). Albany: State University of New York Press.

Tse, L. (2001). *"Why Don't They Learn English?": Separating Fact from Fallacy in the U.S. Language Debate*. New York: Teachers College Press.

University of Wisconsin-Manual on Service Learning (2000). The Morgridge Center: University of Wisconsin-Madison.

Wagner, P. (1994). *A Sociology of Modernity: Liberty and Discipline*. London: Routledge.

Index

accountability, 27–28, 30, 36, 47, 57, 61, 63, 77, 83, 92, 133–134, 187–189, 191, 194, 199–202, 205
   outcome-based, 200
   product-based, 200–202
alchemy, 29, 39–41, 263
Althusser, Louis, 219
Atkinson, Rob, 112, 121
autonomy, 6, 9, 35, 41, 49, 64, 87, 246, 262, 264

Blair, Tony, 1–3, 7, 10, 20, 35, 48–49, 85, 97, 100–101, 103, 106, 113, 134, 190, 215, 225, 233
Bourdieu, Pierre, 44, 217, 255, 264
Britain, 1, 10–11, 20–21, 30–31, 83–85, 95, 97, 101, 103, 106, 213–222, 225–226, 228–230, 232, 234, 238
   Department for Education and Employment (DfEE), 95–96, 100, 103–104, 110, 113–114, 117–118, 121, 130–131, 133–134
   Department for Education and Skills (DfES), 103, 127–128, 133
   Education Reform Act of 1988, 86, 88; *see also* education reform
   General Certificate of Secondary Education (GCSE), 130, 133
   Great Debate, 86, 88, 91
   Institute for Public Policy Research (IPPR), 135, 215, 217
   Labour Party, 1, 20, 22, 85, 215
   Ministry of education, 87, 106–107
   National Curriculum, 83, 86, 111
   North Upton, 96–99, 101–103, 106

Catholic church, 7, 67, 69, 74, 76
centralization, 7–8, 28, 30, 33, 36, 46, 49, 51, 53, 60, 62, 69, 129, 189, 192–193, 195–196, 198–199, 205, 236, 241
Chicago, 21, 34–35, 51, 78, 138, 168, 185, 265
City Technology Colleges (CTCs), 2, 19, 84, 86, 91–94, 99–107
civil society, 2–4, 6, 8, 11, 15, 17, 19, 28, 30–31, 51, 53, 57, 73, 81, 131, 190, 194, 237, 240–242, 246–247, 263, 265
Clinton, Bill, 1, 20, 48–49, 218
Coats, William, 197, 206
colonialism, 52, 213
   neoliberal colonialism, 213
   postcolonial studies, 45
Communism, 8
constructivism, 37, 39, 44–45, 50
crime, 110, 143, 154
cultural practices, 8–9, 16, 28–29, 33, 36, 38–42, 47–48, 50
curriculum enrichment, 123
curriculum standards, 32, 35, 37

decentralization, 7–8, 28, 30, 33, 36, 46, 49, 51, 53, 60, 62, 69, 189, 193, 195–196, 198–199, 205, 236, 241
Deleuze, G., 203–205, 208, 243, 264
democracy, 1, 6–9, 13–15, 17, 19, 21, 27, 29, 34–38, 49, 51, 56, 59–63, 66, 71–73, 75, 105, 107, 115–116, 121, 134–135, 172, 174, 190, 192, 195, 199–200, 205, 214, 216, 219–220, 234, 240, 245, 265
  social democracy, 7, 21, 37, 51, 214, 216, 219–220, 234, 265
  socialist democracy, 66
Derrida, Jacques, 65–66, 71–72, 74–76, 78, 249, 265
deviance, 42–43
Dewey, John, 35, 41, 53
discipline, 40–41, 46, 193, 203–204, 240, 259, 261–262
diversity, 21, 34–35, 46, 78, 92, 105, 114, 141, 156, 164, 166, 223, 226, 257–258, 264
drug prevention, 111

Education Action Forum (EAF), 111, 115–118, 131
ecclesiastic, 58, 64, 67–69
economic disadvantage, 117, 218
  see also poverty
education, 1, 3–5, 7, 9–23, 27–35, 37–40, 42–45, 47, 49–71, 73, 75–79, 81, 83–92, 94–97, 99–107, 109–121, 125–127, 130–140, 143–148, 150–152, 156–158, 160–161, 163–168, 171–202, 204–209, 213–215, 218–241, 243, 245–246, 249–250, 256, 258, 261–265
  access to education, 10–11
  adult education, 61, 148
  educational laws, 63, 66, 227

educational modernization, 58–60
educational priority areas, 90, 100
educational standards, 15, 35, 52, 84, 88, 103
educational success, 177
higher education, 61–62, 67, 171, 256
public funding of education, 89, 91, 175, 197
see also education reform
Education Action Zones (EAZs), 14, 21–22, 84, 86, 94–106, 109–136
EAZ partnerships, 110, 112, 114–115, 117, 131
education reform, 1, 3, 7, 13, 19, 27–29, 32–33, 37, 39, 51, 53, 56, 58–60, 62, 77, 84, 86, 88, 94, 101, 138–140, 143, 146, 187, 189, 196, 200, 202, 205–208, 223, 231, 233, 237, 239–240, 249, 263
Educational Excellence Student Administrative Committee (EESAC), 157, 159–160
educational management, 60, 136
  public management, 30, 135–136
Educational Priority Areas (EPAs), 90
efficiency, 30, 62–63, 179–185, 193, 199–200
Ehrenreich, B., 225, 234
English proficiency, 150, 156, 255
Epstein, Joyce, 151, 177–178, 185, 221, 228, 234
equality, 7, 10–11, 16, 19, 45, 60, 62, 104, 128, 130, 175, 214–216, 220, 224–225, 239, 265
equity, 12–14, 16–18, 27–28, 30, 32, 37, 43, 59–62, 70, 72–73, 130, 135, 175, 179–181, 185, 194, 224, 239

ethnographic studies, 137
  community ethnography, 148, 154
Europe, 2, 6–8, 21–22, 28–29, 44,
    49, 52, 128, 147, 156, 213,
    224, 230, 243
  European Research Network
    Association of Parents and
    Education (ERNAPE), 224
  European Union, 49
  Western Europe, 6, 230, 243
exclusion, 3, 10, 12–16, 19, 21–23,
    42, 48–53, 65–68, 70–73, 75,
    105, 109, 111, 113, 136, 194,
    211, 214, 219, 223, 226–233,
    235, 239–240, 265
experts, 16, 38–39, 41, 85, 101,
    112, 119–120, 128, 176,
    187, 195, 198, 205, 240,
    243, 253–254

faith-based initiatives, 17, 23
families
  family involvement, 144, 177, 237
  female-headed, 218, 223
feminism, 45, 214–218, 220–221,
    233–235
Florida
  A+ Plan for Education, 150,
    156
  Dade County, 140–143, 146,
    148–150, 158, 168
  Miami, 7, 19, 137–138, 140–143,
    145–150, 153–156,
    158–164, 167–168
  Miami-Dade County Public
    Schools (MDCPS),
    140–141, 143, 152,
    160, 168
forums, 116–118, 120
Foucault, Michel, 9, 21, 38, 51,
    53, 65–66, 69, 71, 74, 76,
    78, 192, 208, 240–241, 243,
    249, 253, 256, 265

freedom, 6, 22–23, 48, 53, 59, 142,
    181–182, 184–185, 209, 215,
    241, 243, 262, 265
Friedan, B., 220–221, 234
fundraising, 122, 128, 153

gender, 11–12, 14, 50, 60–62,
    117, 125–126, 175, 181,
    213, 215, 217, 219–224,
    227–228, 230–235, 239,
    244, 251–252, 264
gentrification, 158
Giddens, Anthony, 2, 6, 9–10,
    12, 21–22, 35, 51, 104,
    190, 208, 215–216, 225,
    234, 238, 241, 265
  Runaway World, 216, 234–235
  see also Third Way
globalization, 6–7, 13, 15, 18–19,
    22, 38, 50, 52, 60, 70, 72,
    76–78, 84, 101, 222, 264–265

Halsey, A. H., 90–91
Hanson Industries, 93
Head Start, 5, 11
Holmes Group, 187, 206–207
Hultqvist, K., 33, 51, 265

IBM, 5, 93, 176
immigrants, 34, 142–143, 151–152,
    163, 165, 243
inclusion, 3, 7, 10, 12–16, 18–22,
    37, 42, 50, 53, 56, 61–62,
    66–68, 72–73, 75, 95, 167,
    194, 211, 214, 219, 240, 265
  social inclusion, 3, 7, 10, 12–13,
    15, 20, 95, 214, 240
income, 20, 43, 128, 139–140,
    146–147, 151, 154–156,
    162–163, 165, 168, 177, 184,
    240, 250–251, 255
  lower-income, 162, 165
industrialization, 13

International Monetary Fund (IMF), 56–58, 69–70, 79, 84
investment, 6, 68, 70, 84–85, 93, 101, 112, 115, 121–122, 131, 140, 183–184, 217, 224

Katz, Michael, 174, 186
Kiwanis Club, 145, 153, 160

laissez-faire, 86, 89
laws, 27, 56, 58–59, 63–64, 66–68, 71, 141, 143, 175, 227, 243–244
liberalism
  economic liberalism, 214–215, 224
  social liberalism, 214, 216, 220
Local Education Authorities (LEAs), 87, 89–92, 94–95, 100, 105, 113, 129, 134
localization, 6, 50

market-economy, 70–71
  market forces, 92, 132, 214–215, 233, 235
marriage, 104, 168, 216, 220–221, 228–232
mathematics, 31, 37, 40–41, 50, 52–54, 64, 97, 125, 129, 137–138, 143–144, 153, 168, 183, 186
  international indicators of science and mathematics achievement (TIMMS), 31
mentoring, 123–125, 127, 130, 134, 153, 158, 161, 240, 247–248, 250–252, 254–259, 262
Mexico, 56, 63, 68, 74, 78
  educational reform, 19, 56, 71
  schools, 4, 58

Michigan, 14, 19, 103, 187–188, 190–209, 251
Michigan Partnership, 19, 187–188, 190–193, 195–196, 198–206, 208
military, 29, 44, 46–47, 72
multiculturalism, 7, 34, 145, 150, 156, 175, 204
multi-sector initiatives, 129, 131

National Council for Teachers, 37
National Council of Teachers of Mathematics (NCTM), 144
  see also mathematics
National Science Foundation (NSF), 137, 166, 168
nation-states, 6–7, 22, 27, 52, 243
neighborhoods, 132, 139–140, 142, 151, 166–167
neoliberalism, 2, 7–8, 10–12, 35, 48–49, 70, 72, 105, 190, 213–218, 222–223, 228, 231–232, 241
"No Child Left Behind" Act, 163, 261
normalization, 14, 29, 42–43, 66, 68, 71–72, 192, 194, 198, 202, 243, 245, 261
norms, 5–6, 28–29, 33, 43, 45, 47, 75, 261

parent involvement, 4, 21, 138, 144–145, 163, 165–166, 174, 264
partnerships (375) 1–19, 21–22, 25, 27–30, 32, 35–40, 42, 44, 48–49, 57, 59, 63–66, 68–69, 74, 83–85, 89, 95–97, 99–106, 109–115, 121, 123–124, 127–131, 133, 135–139, 141–142, 144–145, 152, 160–163, 165–166, 169, 171, 175–177, 185, 188, 190, 196, 202, 205, 213–216, 222–226,

228–232, 234, 237–242, 247, 249, 253, 261–263
interschool partnerships, 110, 127
multiagency partnerships, 113
public partnerships, 35, 176
public-private partnerships, 19, 22, 97, 104, 113, 121, 124, 132
signifier partnership, 56, 59–60, 64–65, 71, 73, 75
university-school, 15, 237–239, 249
personal ethics, 38
personal responsibility, 9, 48, 217–219, 227, 231–232
planetspeak, 3–4, 9, 27
Police Athletic League, 153, 158
poverty, 5, 11, 21–23, 42, 45, 57–58, 62–63, 79, 106–107, 114, 143, 146–147, 151, 156, 162, 164, 167, 179, 218, 244, 261
private sector, 5, 19, 59, 63, 65, 83–86, 88–89, 91, 95, 97–99, 102, 110–113, 115, 121–123, 127–128, 131, 133, 135, 171–172, 198, 206, 241
private sector contributions to education, 122–123
professional development, 14, 98, 137, 187, 196, 199–200, 238
Professional Development Schools (PDSs), 187, 196–198, 202, 238
profit, 3, 87, 97, 113, 121, 133, 135, 176, 178, 196, 198, 206–208, 237, 246–247
progress, 13, 15, 23, 27, 30–32, 35, 37, 50, 56, 60–61, 70, 72–73, 75, 118, 129, 131, 134, 156, 163, 178, 189, 221, 225, 238, 240, 243, 264
public service, 5, 59, 63, 70, 76, 84–85, 92, 109, 134, 214–216, 219
see also volunteerism

race, 34, 58, 60, 63, 68–70, 77, 83, 85, 103, 147, 172, 174, 181, 184, 200, 218, 239, 252, 257
racial issues, 5, 143, 156, 175, 244
racism, 11, 21, 163, 233
Ravitch, D., 35–36, 53
reform, *see* education reform
regulation, 3, 14, 87, 179–180, 182, 184, 197, 199
religion, 2, 173–174, 184, 207
rights, 1, 9, 55, 60–61, 63–64, 66–67, 172–173, 190, 245
Rose, Nicholas, 240

salvation narratives, 13, 15, 32, 48–49, 72–74, 237, 249, 251, 257, 263
salvation themes, 13, 15, 27–30, 36–37, 47, 49
Saussure, F. de, 74, 78
scholarships, 14–15, 41, 145, 153, 160, 162–163, 186
schools, 1, 3–8, 10–23, 27–44, 46–47, 49–50, 52–54, 56–65, 67–70, 74, 76, 83–107, 110–116, 119–135, 137–142, 144–169, 171–190, 192–209, 214–215, 218, 220–228, 230–240, 242–244, 246–266
charter schools, 32–33, 175, 182, 186, 189, 196–199, 202, 206, 208, 236
private schools, 12, 57, 59, 63, 74, 76, 88, 155, 159–160, 171–172, 174–176, 180–181, 183, 197
public schools, 18, 63–64, 69, 76, 88, 103, 112, 140–142, 150, 159, 168, 171, 173–176, 181–184, 186–187, 194, 196, 206–208, 234, 244, 250, 253–255

school choice, 33–34, 104, 186, 235
  see also vouchers
school improvement plan, 201–202, 204, 207
school management, 28, 30–31, 57, 193, 198
  government-only management, 62
school reform, *see* education reform
Science Engineering Communication Math Enhancement (SECME), 144, 153
sex education, 11, 20, 50, 77, 213–217, 219–223, 225, 228–235
  abstinence-only, 229
SHAPE Tutoring Project, 239, 242, 249, 264
social justice, 30, 70, 103–104, 135–136, 232–233
social policy, 13, 21, 23, 135, 235
social science, 29, 34, 44–45, 213, 217, 233, 235
social welfare, 35, 48, 60, 195, 218–219, 228, 231, 233–235, 244, 254
  AFDC (Aid to Families with Dependent Children), 218
  Temporary Aid for Needy Families (TANF), 218
  welfare reform, 9, 218, 228, 232
solidarity, 23, 35, 58, 60–61, 64
standardized testing, 189, 205, 254
suburbs, 43, 94, 220–221
Sweden, 32–34, 37, 49–53

teachers, 5, 13, 17, 19–23, 28–29, 31, 33, 36–44, 46–47, 50–54, 57, 59–64, 78, 87, 90–92, 96–98, 101–103, 105–106, 111, 113–114, 116–117, 120, 122–123, 125–129, 131, 135–136, 144–145, 149, 151–152, 157, 160, 163, 166, 180, 186–187, 191, 193, 197, 199, 201, 204, 206–207, 209, 227, 238–240, 244–249, 251–262, 264–266
teaching, 28, 31–33, 39–41, 43, 49–51, 53, 57, 63, 87, 90, 105–107, 116, 124, 137, 157, 173, 181, 197, 207–208, 238, 245, 251, 254–255, 257
Teaching Quality Assurance (TQA), 31–32
Technical and Vocational Educational Initiative (TVEI), 92, 105
textbooks, 41, 87, 176
Thatcher, Margaret, 1, 21, 48–49, 83, 85, 91, 105–106, 214, 216
Third Way, 2, 11–12, 15–16, 20–23, 35, 48, 51, 85, 97, 100, 103–104, 106, 134–136, 189–194, 197–198, 200, 205, 208–209, 215, 225, 231, 233–234, 238, 241–242, 265
  see also Giddens, Anthony
Thompson, Edward, 102, 107
training, 5, 9–11, 31, 33, 46, 52, 61, 67–68, 85, 106, 110–111, 116, 148, 177, 203, 238, 249, 256
training and enterprise council, 110–111
tutoring, 20, 63, 89, 91, 97, 111, 157, 176–177, 226, 237–240, 242, 246–262, 264

UNESCO, 56–58, 65, 69–70, 79
United States, 1, 7–8, 33, 44–45, 49, 137, 142–143, 151, 163, 168, 182, 188–190, 195, 213–215, 217–218, 220–222, 224–225, 228–230, 232, 235, 237–238, 243–244, 247, 252, 261–262, 266
universalization, 66, 70
universities, 15, 19–23, 31, 49–53, 61, 77–78, 87, 89–90, 103–107, 134–135, 137, 157,

161, 167–168, 185–187,
191–192, 196, 198–200,
205–209, 215, 233–240, 246,
249–261, 264–266
urbanization, 13, 62
Urban Systemic Initiative (USI), 138,
141, 143–144, 166

voluntary agencies, 2, 86, 95, 127
volunteerism, 15, 59, 61, 64, 97,
123, 125–126, 145–146, 153,
158, 162–165, 176, 178, 201,
237, 241–242, 244–248, 256,
258, 263
Citizen Conservation Corps
(CCC), 244
Peace Corps, 245
vouchers, 14, 32, 172, 175, 178–186,
189, 197, 224, 236

Walkerdine, V., 50
welfare, *see* social welfare
World Bank (WB), 56–58, 65,
69–70

GPSR Compliance
The European Union's (EU) General Product Safety Regulation (GPSR) is a set of rules that requires consumer products to be safe and our obligations to ensure this.

If you have any concerns about our products, you can contact us on

ProductSafety@springernature.com

In case Publisher is established outside the EU, the EU authorized representative is:

Springer Nature Customer Service Center GmbH
Europaplatz 3
69115 Heidelberg, Germany

www.ingramcontent.com/pod-product-compliance
Lightning Source LLC
LaVergne TN
LVHW011807060526
838200LV00053B/3686